Elementary
Number Theory

The Random House/Birkhäuser Mathematics Series:

INVITATION TO COMPLEX ANALYSIS
  by R.P. Boas

BRIDGE TO ABSTRACT MATHEMATICS: MATHEMATICAL
PROOF AND STRUCTURES
  by Ronald P. Morash
ELEMENTARY NUMBER THEORY
  by Charles Vanden Eynden
INTRODUCTION TO ABSTRACT ALGEBRA
  by Elbert Walker

Elementary Number Theory

Charles Vanden Eynden

Illinois State University

The Random House/Birkhäuser Mathematics Series

Random House
New York

Permissions/Acknowledgments

Page 61: Photo from D. N. Lehmer, "Machine Performs Difficult Mathematical Calculations." *News Service Bulletin*, 1939, Vol. III, No. 3. Reprinted by permission of the Carnegie Institution of Washington.

Page 87: "A 65,000-Digit Number" from *The Washington Post*, September 23, 1985, p. A7. Copyright © 1985 by The Washington Post. Reprinted by permission.

Page 87: "Prime Number Record Broken" from the London *Times*, November 17, 1978, p. 9. Reprinted by permission of Agence France-Presse.

Page 144: "Opening the 'Trapdoor Knapsack'" from *Time* Magazine, October 25, 1982, p. 88. Copyright © 1982 by Time Inc. All rights reserved. Reprinted by permission from *Time*. Cartoon reprinted by permission of Larry Ross.

Page 145: "N. M. Scientist Cracks Code" from *The Boston Globe*, November 6, 1982. Reprinted courtesy of The Boston Globe.

Page 224: Photo of G. H. Hardy from *A Mathematician's Apology*, by G. H. Hardy. Reprinted by permission of the publisher, Cambridge University Press. Photo of S. Ramanujan from *Ramanujan*, by G. H. Hardy. Reprinted by permission of the publisher, Cambridge University Press.

First Edition
987654321
Copyright © 1987 by Random House, Inc.

Library of Congress Cataloging in Publication Data

Vanden Eynden, Charles, 1936–
   Elementary number theory.

   Bibliography: p.
   Includes index.
   1. Numbers, Theory of.   I. Title.
QA241.V26   1987   512'.72   86–21929
ISBN 0–394–35359–5

Manufactured in the United States of America by Halliday Lithograph, Hanover, Ma.
Jacket design by David Lindroth

To my parents,
Lawrence and Sophia
Vanden Eynden

Preface

This book was written with four goals in mind:

1. To be easy for the student to read and easy for the professor to teach from.
2. To be suitable for a wide variety of audiences.
3. To involve the student actively in the construction of proofs.
4. To develop in the student a self-confidence with respect to mathematics that will foster interest and success.

To achieve the above goals I have set off from the text definitions and examples, as well as theorems, so that they are easy to find and so that the central topics of each section are apparent. In addition, each section is followed by a set of problems divided into three parts. The problems in part A are generally straightforward calculations using techniques covered in the section. Problems in part B include calculations needing some ingenuity or new ideas and standard proofs. Part C problems call for harder proofs. (Of course, the classification of the problems, especially those in parts B and C, calls for a judgment with which the reader may sometimes disagree.) Also, to give a taste of how advances in mathematics really are made, a large number of true-false problems are given. These foster experimentation and a healthy skepticism.

I believe number theory is an ideal subject for introducing the craft of proof construction, by reason of its concreteness and familiarity. However, proofs cause great anxiety in many students; there is nothing safe about finding a proof, the way doing a calculus problem is safe (one merely learns the technique, then applies it). Anxiety with respect to proofs makes success less likely in most cases. Therefore, the number of computational problems in this book exceeds that of most number theory texts. A substantial portion of problems of a computational or algorithmic nature provide the student with a floor of activities he or she can feel confident about. Computational problems also lead to understanding of the theory behind them. Moreover, the large supply of easier problems also enables the professor to make an assignment every day, even when not much is covered in class.

To further reduce anxiety about proofs, the ideas behind a proof are explained whenever possible. The student is shown that theorems and proofs do not spring from the pens of mathematicians in a perfect form, ready to be printed in books, but rather are the result of trial and error and evolutionary improvement just as most high-level human accomplishments are. Fortunately, number theory has a rich history, and many historical details are included, both for their intrinsic interest and to show that mathematics is an activity of people.

Historically in mathematics a problem arises, and then (sometimes) it is solved. In textbooks, however, it is usual to give the solution, followed by applications. This order of presentation is compact and logical, but the historical sequence is generally better pedagogy. Whenever possible, I have tried to set up theorems as solutions to naturally arising problems or conjectures. Occasionally this approach requires introducing a concept before it can be developed fully; an example is the function $d(n)$ in Section 2.1.

I have aimed for a writing style that is informal yet to the point. If a theorem needs a few words to put it in context or an example to illustrate either its proof or application, it is provided; but I have tried to avoid saying what is obvious. Section headings and other titles have been chosen to be informative rather than clever. Over the years I have come to the sad realization that many students don't read their expensive textbooks anyway; they try to do the problems and look back for a similar example if they get stuck. (This syndrome may be acquired from calculus books so monumental that no one could be expected to read every word.) My aim has been to talk straight to those who will read the text and provide a convenient organization for the others.

Several topics receive special emphasis in this book. A very short introduction to logic and mathematical proof is given in Chapter 0. A section on mathematical induction is provided in Chapter 1 in order to get across the idea that induction is more than an algorithm for proving summation formulas. In addition, the use of the sigma and pi notations for sums and products is treated at length, especially with respect to the way these notations are used in number theory. Chapter 6 opens with a section on countability as it applies to the rational and real numbers. Although this is somewhat unusual in a number theory text, I find this material puts continued fractions in a richer context, and undergraduates are sometimes unfamiliar with it. Later sections do not depend upon this one. I have also included a section introducing the RSA method of public key cryptography, because students seem to enjoy this application of number theory.

Since almost all students now have calculators, tricks for using them in number-theoretic algorithms are included. Occasionally a BASIC program is given to illustrate a more complex computation. Of course, these discussions may be covered or ignored, as the instructor desires.

Theorems have been numbered consecutively in chapters, so that Theorem 1.3 is the third numbered theorem in Chapter 1. Although such numbering is sometimes convenient because of later references, students can hardly be expected to remember these numbers. Thus whenever possible I have used names for theorems (for example, the Euclidean Algorithm), and theorems not referred to later are not numbered. Sometimes an important theorem is not numbered because a more general version is coming up.

Depending on the speed with which topics are covered and what is expected from the student, it is possible to use this book for a wide range of courses, from a number-theory-appreciation treatment emphasizing history and type A problems to a proof-oriented course preparing the student for graduate courses. A minimal course would probably include the first four chapters, but even so, Sections 3.6 and 4.5 and some topics (such as Fibonacci numbers, dividing exactly, and amicable, abundant and deficient numbers), could easily be skipped. The last three chapters are largely independent of each other, except that in Chapter 7 the treatment of sums of four squares depends on Section 5.3 and the presentation of Pell's equation depends on Section 6.3.

Acknowledgments

This book has been improved by the corrections and suggestions of many conscientious readers. The following reviewers can take credit for much of what is good in the book and blame for none of my stubborn idiosyncrasies:

Gerald L. Alexanderson, University of Santa Clara
Bruce C. Berndt, University of Illinois
Roger Entringer, University of New Mexico
Joel K. Haack, Oklahoma State University
Neal Hart, Sam Houston State University
Padmini T. Joshi, Ball State University
Carole B. Lacampagne, University of Michigan—Flint
Arne Magnus, Colorado State University
Paul J. McCarthy, University of Kansas
Charles J. Parry, Virginia Polytechnic Institute and State University
William H. Richardson, Wichita State University
Daniel H. Saracino, Colgate University
Lawrence C. Washington, University of Maryland
William A. Webb, Washington State University

This list attests to the care given this book by John R. Martindale, Judith Kromm, and others at Random House, in addition to many creative ideas for its improvement.

Finally, it has been my pleasure to benefit from the lectures and writings of many fine number theorists, but especially those of Ivan Niven.

Contents

Elementary
Number Theory

Chapter 0

What Is Number Theory?

No completely satisfactory definition has ever been given, but roughly, number theory is the mathematical treatment of questions related to the integers, that is, the numbers $0, 1, -1, 2, -2, \ldots$. Investigating such questions has often led mathematicians into other fields, however, such as complex numbers, geometry, and abstract algebra.

Number theory may very well be the best subject for a student trying to learn what constitutes a mathematical proof, and to begin to construct proofs. It combines the advantages of concreteness and familiarity. The integers possess a concreteness, a realness, that more abstract concepts (vector spaces, for example) do not. This concreteness allows us to form hypotheses about the integers and test them by actual experiment.

Of course the integers are familiar to us from our earliest introduction to arithmetic. We have manipulated them thousands of times, and have formed an intuitive sense of many of their laws. This intuition carries some danger with it, because it may be hard to see the necessity for a proof to verify patterns that we have confirmed by instances over and over again since grade school. Although intuition can be of immense help in mathematics, no accumulation of special cases is sufficient to prove a general proposition.

This book contains a large quantity of exercises, mainly of two types: computations and proofs. There are many more computational problems than in most number theory texts. The student who finds a particular section hard to navigate is advised to self-administer large doses of these computational problems; they lead to understanding.

Proofs are another matter. A calculus student may feel he or she should be able to do any assigned problem, by applying the standard methods of the subject. On any particular day, however, some person may or may not be able to construct a proof of a given mathematical fact. There are no infallible methods. If there were, the proof would be unnecessary; one could merely say "Apply method A." Thus no one should expect to plow through a set of proof problems like a list of differentiation problems, knocking them off one after the other. A success rate of 100% is too much to expect. Sometimes the right idea comes at once, sometimes after an hour's work, sometimes after a night's sleep, sometimes never.

Most assertions to be proved are **implications**, that is, statements of the form "if P, then Q," where P and Q are themselves statements. The statement P is called the **hypothesis** and Q is called the **conclusion**. An example (talking about real numbers) is:

> S: If $x > 4$,   then $x > 2$.

Here P is the statement "$x > 4$," Q is the statement "$x > 2$," and the implication S is, of course, true. Notice that "if P, then Q" is not the same as "if Q, then P"; and one of these implications may be true and the other false. In our example the second statement would be the assertion:

> S': If $x > 2$,   then $x > 4$.

Of course S' is false, as the **counterexample** $x = 3$ shows. The statement "if Q, then P" is called the **converse** of the statement "if P, then Q."

The implication "if P, then Q" is considered to be true unless an instance exists when P is true but Q is false. Suppose I make the statement to my daughters "If it is sunny Saturday, then we will go to the zoo." My statement is truthful in each of the following cases:

a. Saturday is sunny and we go to the zoo.
b. Saturday is not sunny and we do not go to the zoo.
c. Saturday is not sunny but we go to the zoo anyway.

I have lied only if Saturday is sunny but we do not go to the zoo. The meaning of "if P, then Q" is summarized in the following truth table.

P	Q	if P, then Q
true	true	true
true	false	false
false	true	true
false	false	true

There are many sets of True-False exercises in this book. The idea is to prove those statements that are true and disprove the others. The disproofs ordinarily take the form of a counterexample. The more concrete and definite the counterexample is, the better. For example, consider the statement

> If $a \geq b$, then $ac \geq bc$.

This could be disposed of by

> FALSE. Let $a = 2$, $b = 1$, and $c = -1$. Then $a = 2 \geq 1 = b$, but $ac = -2$ and $bc = -1$. Thus $ac \geq bc$ is false.

This is superior to

FALSE. If $a \geq b$ and c is negative, then $ac < bc$.

The second disproof is inferior in two respects. First, it is more general than it has to be, invoking a theorem about inequalities when all that is really needed is that $-2 < -1$. Second, it is itself false. What if $a = b = 0$ and $c = -1$? Then $a \geq b$ and c is negative, yet $ac < bc$ is false. True enough, the second disproof is valid except in the special case that $a = b$. But that is enough to invalidate it. Mathematics is a very picky subject.

It should be mentioned that a purely numerical counterexample does not always suffice to disprove a statement. Consider

There exists x such that $x > y$ for all y.

It is not good enough to say

FALSE. Let any $x = 100$. Then if $y = 200$, $x > y$ is false.

After all, nobody said x was 100. It must be shown that *no* x satisfies the statement.

FALSE. Let any x be given. Set $y = x + 1$. Then $x > y$ is false.

The above is a satisfactory disproof.

Sometimes when you are trying to prove something and are stumped, it helps to look for a counterexample. If you can see why you can't find one, you are on your way to a proof.

It is a ground rule that in proving an exercise any previous exercise or result of the text may be assumed unless the contrary is expressly stated.

Some problems and many theorems contain the words "if and only if." Proving that P if and only if Q (where P and Q are statements) usually breaks into two parts:

1. P if Q (i.e., whenever Q is true, so is P), and
2. P only if Q (i.e., whenever P is true, so is Q).

As an example, consider the familiar statement:

If A, B, and C are real numbers with $B^2 - 4AC > 0$ and $A \neq 0$, then $Ax^2 + Bx + C = 0$ if and only if x is

$$\frac{-B + \sqrt{B^2 - 4AC}}{2A} \quad \text{or} \quad \frac{-B - \sqrt{B^2 - 4AC}}{2A}.$$

The "if" part of the statement says that each of the two numbers given satisfies the quadratic equation, and could be proved by direct substitution. The "only if" part says these are the only solutions, and is usually proved by completing the square on the original equation.

Sometimes we may say that a proof is "indirect," or a "proof by contradiction." This simply means that the scheme of the proof is to assume that the statement to be proved is false. Then contradictory statements are proved. From this we conclude that the statement to be proved must be true. As an example we offer (with a few comments enclosed in brackets) the standard proof that $\sqrt{2}$ is irrational (that is, not the quotient of two integers).

THEOREM   The number $\sqrt{2}$ is irrational. [The words "the number" were inserted just to avoid starting a sentence with a symbol.]

Proof   We will give an indirect proof. [It is always wise to let the reader in on the type of proof we have in mind.] Suppose $\sqrt{2}$ is rational. [Here is where we assume the statement to be proved is false.] Suppose

$$\sqrt{2} = \frac{a}{b},$$

where a and b are integers and the fraction a/b is in lowest terms. [When the new symbols a and b are introduced we are careful to explain what they represent.]
  Squaring both sides leads to

$$2b^2 = a^2.$$

From this equation we see that a^2 is even. Thus a must be even, for if $a = 2k + 1$, then

$$a^2 = (2k + 1)^2 = 4k^2 + 4k + 1$$
$$= 2(2k^2 + 2k) + 1,$$

which clearly is odd. [This part of the proof illustrates one of the hardest questions involved in presenting an argument, namely, how much detail to include. If this proof were given later in this course it would probably be reasonable to assume the reader would know that the square of an odd integer had to be odd. But since we are only in Chapter 0 we have included the explanation beginning with "for if $a = 2k + 1$." It is always better to err in the direction of giving too much detail.]
  Since a is even, let $a = 2c$, where c is an integer. [The new symbol c is explained.] Then $2b^2 = 4c^2$, and so

$$b^2 = 2c^2.$$

[Have we been too optimistic in assuming the reader will see we substituted $a = 2c$ into the equation $2b^2 = a^2$?]
  Thus we see that b^2 is even, and so b must be even, as above. [No sense in repeating the argument we just made for a.]

But this contradicts the assumption that a/b was in lowest terms, since we have seen that both a and b must be even. This contradiction proves that $\sqrt{2}$ must be irrational.

Note that in the above proof we had to indicate specifically that the variables a, b, and c were integers. We will save time in the future by adopting the following convention:

CONVENTION   Since the integers will be our main concern, we will make the convention that all lower case italics letters $(a, b, c, \ldots, x, y, z)$ will denote integers.

Number theory is quite an old subject, and some of the theorems in this book may be found in the writings of Euclid, who lived about 300 years before Christ. It might be thought that all the natural questions that arise about the integers have been solved for a long time, but this is not the case. The field abounds in problems that can be stated simply but that no one has been able to solve for centuries. We will list a few of these here; later in the book historical details of some of them will be given.

A few simple definitions are needed. These will be given quickly here and repeated later when the theory is developed in a more systematic way. We say that the integer a is a **divisor** of the integer b if b/a is an integer; for example, 3 is a divisor of 15, but 3 is not a divisor of 16. We say that the integer p is **prime** if $p > 1$ and the only positive divisors of p are 1 and p itself; the first three primes are 2, 3, and 5.

Unanswered Questions

a. Most people are familiar with solutions to $x^2 + y^2 = z^2$, for example $x = 3$, $y = 4$, and $z = 5$. But for $n > 2$ does $x^n + y^n = z^n$ ever have a solution in positive integers x, y, and z?   [Fermat's Last Theorem]

b. Many examples of pairs of primes differing by 2 can be found, such as 3 and 5, 5 and 7, 11 and 13, and 17 and 19. Are there infinitely many of these?   [The Twin Prime Conjecture]

c. It seems to be possible to write each even number greater than 2 as the sum of 2 primes. For example, $4 = 2 + 2$, $6 = 3 + 3$, $8 = 3 + 5$, and $10 = 3 + 7$. Can this always be done?   [The Goldbach Conjecture]

d. Are there infinitely many primes of the form $n^2 + 1$, such as $1^2 + 1 = 2$, $2^2 + 1 = 5$, $4^2 + 1 = 17$, and $6^2 + 1 = 37$?

e. Are there infinitely many primes of the form $2^n - 1$, such as $2^2 - 1 = 3$, $2^3 - 1 = 7$, and $2^5 - 1 = 31$?   [Mersenne Primes]

f. Are there infinitely many primes of the form $2^n + 1$, such as $2^1 + 1 = 3$, $2^2 + 1 = 5$, and $2^4 + 1 = 17$?   [Fermat Primes]

g. An integer is said to be **perfect** if it is the sum of its positive divisors other than itself. An example is $6 = 1 + 2 + 3$. Are there infinitely many perfect numbers?

h. Are there infinitely many odd perfect numbers? Is there even one?

i. The decimal expansion of $\frac{1}{7}$ is $0.142857142857\ldots$, where the 6 digits 142857 repeat forever. Are there infinitely many integers n such that the smallest set of repeating digits in the decimal expansion of $1/n$ has length $n - 1$?

Lest the reader be left with too pessimistic a view of number theory we will list several results that *can* be proved. Most of these will be found in this book, but a few have proofs too difficult to include.

a. The equation $x^2 + y^2 = z^2$ has infinitely many solutions in positive integers with no common divisor greater than 1. (See problem (a) above.) In fact in some sense the solutions to this equation have been completely characterized; see Section 7.1. [Pythagorean Triples]

b. There are infinitely many primes. [Euclid]

c. In fact the series $\frac{1}{2} + \frac{1}{3} + \frac{1}{5} + \cdots$ with the reciprocals of the primes as terms diverges. [Euler]

d. On the other hand, if we restrict the series in (c) to primes differing by 2 from another prime (see problem (b)), the series $\frac{1}{3} + \frac{1}{5} + \frac{1}{7} + \frac{1}{11} + \frac{1}{13} + \cdots$ converges. (Of course this may be a finite sum.) [Brun's Theorem]

e. If a and b are positive integers with no common divisor greater than 1, then the arithmetic progression $a, a + b, a + 2b, \ldots$ contains infinitely many primes. An example, with $a = 3$ and $b = 10$, is that the sequence 3, 13, 23, 33, 43, $\ldots$ contains infinitely many primes. [Dirichlet's Theorem]

f. If $2^n - 1$ is prime, then n itself must be prime. (See problem (e).)

g. If $2^n + 1$ is prime, $n > 0$, then n must be a power of 2. (See problem (f).)

h. If n is an *even* perfect number, then n has the form $2^{k-1}(2^k - 1)$, where $2^k - 1$ is prime. Conversely, if $2^k - 1$ is prime, then $2^{k-1}(2^k - 1)$ is perfect. (See problem (g).) [Euler, Euclid]

i. If the smallest set of repeating digits of $1/n$ has length $n - 1$, then n is prime. (See problem (i).)

Even though most of the questions in the first list we gave have been unanswered for many years, even centuries, progress continues to be made on some of them. We will consider the history of problem (c), which asks whether each even integer greater than 2 can be written as the sum of two primes. On the basis of numerical evidence the German-born Christian Goldbach conjectured that the answer was yes in a letter to the Swiss mathematician Euler in 1742; and this guess is now known as the **Goldbach Conjecture**. It was first published by the English mathematician Edward Waring in 1770, but little progress was made on proving the conjecture until the twentieth century.

Partial results have been mainly of two types: theorems saying that every sufficiently large even number is the sum of at most k primes, where proving that one can take $k = 2$ is the ultimate goal, and theorems saying that every sufficiently large even number is the sum of two "nearprimes," that is, numbers with a limited number of prime factors.

The best-known result in the first direction is the 1937 proof by the Russian Vinogradov that every sufficiently large positive integer is the sum of at most 4 primes. The words "sufficiently large" deserve more explanation. What is meant is that there exists an integer N such that if n is any even integer with $n > N$, then n can be expressed as the sum of 4 primes. Thus any even integers not the sum of 4 primes all lie in some known, bounded interval, and there is the possibility of checking the even numbers in this interval by direct computation. In Vinogradov's original paper the value of N was about

$$10^{10^{10^{17.86}}},$$

a number large enough to discourage even a computer, but later mathematicians were able to prove the theorem for smaller values of N.

Actually Vinogradov proved that every sufficiently large *odd* integer is the sum of at most 3 primes. The result for even integers is an easy consequence of this, since if n is even, then $n - 3$ is odd. Then from $n - 3 = p_1 + p_2 + p_3$, with p_1, p_2, and p_3 prime, follows $n = 3 + p_1 + p_2 + p_3$, and so n is the sum of 4 primes. Vinogradov's theorem had been preceded by the 1930 proof by Schnirelmann that an integer k exists such that every integer can be written as the sum of at most k primes, but Schnirelmann's value of k was quite large.

Goldbach-type theorems involving the sum of two "nearprimes" go back to Viggo Brun, who proved in 1919 that every sufficiently large even number can be written as the sum of two integers, each of which is the product of at most 9 primes. (In these results a prime is counted in a product according to its exponent, so 12 would be considered a product of 3 primes: 2, 2, and 3.) By 1940 A. Buchstab had reduced the number 9 to 4. In 1948 A. Rényi demonstrated the existence of a number k such that every large even number is the sum of a prime and an integer having at most k prime factors, and by 1963 the mathematician Pan had reduced k to 4. This improved Selberg's 1950 result that every large even number could be expressed as $P + Q$, where PQ had at most 5 prime factors. In 1965 Buchstab proved the Rényi theorem with $k = 3$. The best theorem to date along these lines is that of the Chinese mathematician Chen, announced in 1966 and published in 1973, that every sufficiently large even integer is the sum of a prime and a number that is the product of at most 2 primes. If 2 could be replaced by 1 the Goldbach conjecture would be essentially proved.

This little history indicates the international nature of mathematics. A conjecture made by a German who lived in Russia in a letter to a Swiss and published by an Englishman has been worked on by mathematicians of many nations, with the last word so far belonging to a Chinese.

Probably no area of mathematics has been studied more for the sheer enjoyment of its ideas and problems than the theory of numbers. The author hopes that *you*, the reader, will discover some of the beauty and ingenuity that past generations of number theorists, amateur and professional, have found in the subject.

Chapter 1

Divisibility

1.1 The Greatest Common Divisor and Least Common Multiple

The sum, difference, and product of two integers is always an integer, but the quotient may not be. We will start by investigating this most troublesome operation: division of integers.

DEFINITIONS   divides, divisor, multiple, greatest common divisor, least common multiple

We say a **divides** b and write $a|b$ in case b/a is an integer. In this case we also say that a is a **divisor** of b and that b is a **multiple** of a. If $a|b$ is false we write $a \nmid b$.

Examples

$$3|6, \quad -3|6, \quad 2|-6, \quad 3|0, \quad 5|5, \quad 2 \nmid 5, \quad 8 \nmid -4.$$

We say d is the **greatest common divisor** (gcd) of a and b in case d is the largest of all the integers dividing both a and b, and we write $d = (a, b)$.

Example   Let $a = 4$ and $b = 6$.

Divisors of 4: $1, -1, 2, -2, 4, -4$.
Divisor of 6: $1, -1, 2, -2, 3, -3, 6, -6$.
Common divisors of 4 and 6: $1, -1, 2, -2$.

Thus the greatest common divisor of 4 and 6 is $2 = (4, 6)$. Likewise (since the divisors are the same) $(-4, 6) = (4, -6) = (-4, -6) = 2$.

We say m is the **least common multiple** (lcm) of a and b in case m is the smallest of all the positive integers that are multiples of both a and b. We write $m = [a, b]$.

Example   Let $a = 4$ and $b = 6$.

> Positive multiples of 4:   4, 8, 12, 16, 20, 24, 28, ...
> Positive multiples of 6:   6, 12, 18, 24, 30, ...
> Common positive multiples of 4 and 6:   12, 24, ...

Thus the least common multiple of 4 and 6 is $12 = [4, 6]$. Likewise (since the multiples are the same), $[-4, 6] = [4, -6] = [-4, -6] = 12$.

The concept of least common multiple should be familiar from arithmetic. To add $\frac{3}{4}$ and $\frac{1}{6}$ we write

$$\frac{3}{4} + \frac{1}{6} = \frac{9}{12} + \frac{2}{12} = \frac{11}{12}.$$

The 4ths and 6th are converted to 12ths, 12 being the least common multiple of 4 and 6.

The following statements are nearly obvious from the definitions and will be used freely in what follows.

a. The following are equivalent: $a|b$, $-a|b$, $a|-b$, $-a|-b$.
b. If (a, b) is defined, then $(a, b) = (b, a) = (-a, b) = (a, -b) = (-a, -b)$.
c. If $[a, b]$ is defined, then $[a, b] = [b, a] = [-a, b] = [a, -b] = [-a, -b]$.

Some Simple Divisibility Theorems

The proofs of Theorems 1.1 and 1.2 follow quite directly from the definition of dividing, and the reader may want to try devising his or her own proofs before reading the ones given.

THEOREM 1.1   If $a|b$ and $b|c$, then $a|c$.

Proof   By the definition of $a|b$, we know that b/a is an integer. Likewise c/b is an integer. Then so is the product of these quotients an integer, namely

$$\frac{b}{a} \cdot \frac{c}{b} = \frac{c}{a}.$$

This means that $a|c$.

Many authors define $a|b$ to mean that $b = ac$ for some integer c. This definition is sometimes easier to use than the one given at the start of this section.

The two definitions are equivalent except for the uninteresting case $a = b = 0$. (The relation $0|0$ is false using our first definition but true for the alternative one.) We will use whichever definition is more convenient in the future. The proof given for the next theorem uses the second definition.

THEOREM 1.2   If $a|b$ and $a|c$, then $a|bx + cy$ for any integers x and y.

Proof   Since $a|b$ there exists an integer r such that $b = ra$. Likewise there exists s such that $c = sa$. Then

$$bx + cy = (ra)x + (sa)y = a(rx + sy).$$

This means that $a|bx + cy$.

A Property of the Least Common Multiple

Recall that $[a, b]$ is defined to be the smallest positive common multiple of a and b. Let us compute $[15, 10]$ from the definition.

Positive multiples of 15:   15, 30, 45, 60, 75, 90, 105, ...
Positive multiples of 10:   10, 20, 30, 40, 50, 60, ...
Common positive multiples of 15 and 10:   30, 60, 90, 120, ...

We see that $[15, 10] = 30$, the smallest element in the third list. Notice that the other common positive multiples of 15 and 10 appear to be exactly the positive multiples of 30, the least common multiple. This suggests the following result.

THEOREM   If $a|m$ and $b|m$, then $[a, b]|m$.

Proof   It is not quite as easy to see how to form a proof here as in the first two theorems. We wish to show that $m/[a, b]$ is an integer, but it is not clear how this follows from the fact that m/a and m/b are integers.

We will try an indirect proof (see Chapter 0). That is, we will assume that $m/[a, b]$ is *not* an integer, and try to derive a contradiction from this.

Certainly if $m/[a, b]$ is not an integer, then for some integer q and real number ε

$$\frac{m}{[a, b]} = q + \varepsilon, \qquad \text{where} \quad 0 < \varepsilon < 1.$$

Here q is simply the largest integer less than $m/[a, b]$. (We use a Greek letter for ε since *Roman* letters are reserved for integers.) Multiplying through by $[a, b]$ gives

$$m = q[a, b] + \varepsilon[a, b], \qquad \text{and} \qquad 0 < \varepsilon[a, b] < [a, b].$$

(Multiplying the inequalities is justified, since by definition $[a, b] > 0$.)

Notice that $\varepsilon[a, b]$ is the integer $m - q[a, b]$. Let us denote it by r. Then

$$m = q[a, b] + r, \quad \text{and} \quad 0 < r < [a, b].$$

Now $r = m - q[a, b]$. By hypothesis $a|m$, and $a|[a, b]$ by the definition of the lcm. Thus $a|r$ by Theorem 1.2 (with $x = 1$ and $y = -q$). The same argument shows that $b|r$. We see that r is a common multiple of a and b. But the inequality $0 < r < [a, b]$ contradicts the fact that $[a, b]$ is the *least* positive common multiple of a and b.

Since the assumption that $[a, b]$ does not divide m leads to a contradiction, we conclude that $[a, b]$ divides m.

Writing Proofs

The proofs of three theorems are given in this section. From them we may draw some simple lessons about how proofs are expressed.

A proof should be written in sentences. These sentences are written in the English language, even though some technical words or symbols may be used. They should start with capital letters (so a sentence should not begin with a symbol), and end with periods. Other conventional punctuation should be used as appropriate.

The most important thing to think about when writing up a proof is who is going to read it. The mathematician Ivan Niven has said, "A proof is something that convinces somebody." It is a communication between humans. It is not sufficient that the writer of the proof believes it; enough detail must be given so that the reader will also understand and believe. The burden of making oneself understood rests on the writer of the proof.

PROBLEMS FOR SECTION 1.1

1. True or false? $7|203$, $16|-1000$, $-6|3$.
2. True or false? $75|3000$, $71|0$, $0|-12$.
3. Find $d > 0$ such that $d|18$, $d \nmid 12$, and $36/d \nmid 10$.
4. Find $d > 0$ such that $d \nmid 1000$, $5|d$, $d|60$, and $d/2|75$.
5. Find $(51, 34)$ and $[51, 34]$.
6. Find $(16, 81)$ and $[16, 81]$.
7. Find $(360, -2250)$ and $[360, -2250]$.
8. Find $(434, 589)$ and $[434, 589]$.
9. Find all $d > 0$ such that $18|d$ and $d|216$.
10. Find all $d > 0$ such that $20|d$ and $d|300$.
11. What are all the divisors of 24?
12. What are all the divisors of 30?
13. What are the multiples of 4 between -25 and 25?
14. What are the multiples of 5 between -42 and 42?

15. Make a table showing b, (a, b), $[a, b]$, and $(a, b)[a, b]$ for $a = 8$ and b running from 1 to 9.
16. Make a table as in problem 15 for $a = 9$ and b running from 1 to 10.

17. For what integers a is $1 \mid a$ true?
18. For what integers a is $a \mid 0$ true?
19. For what integers a is $a \mid b$ true for all integers b?
20. Why does the word "positive" appear in the definition of $[a, b]$ but not in the definition of (a, b)?

True-False. In the next nine problems tell which statements are true and give counterexamples to those that are false. Assume a, b, and c are arbitrary nonzero integers.

21. If $ab > 0$, then $[a, b] \le ab$.
22. If $c \mid a$ and $c \mid b$, then $[a, b] \le ab/c$.
23. If $(a, b) = 1$ and $(a, c) = 1$, then $(b, c) = 1$.
24. $(a, b) \mid [a, b]$.   25. If $b \mid c$, then $(a, b) \le (a, c)$.
26. If $b \mid c$, then $[a, b] \le [a, c]$.   27. If $a \mid b$ and $b \mid c$, then $a \mid c$.
28. $(ac, bc) = c(a, b)$.   29. $(ac, bc) = |c|(a, b)$.

30. Show that (a, b) is defined if and only if a and b are not both 0.
31. Show that $[a, b]$ is defined if and only if neither a nor b is 0.
32. Show that neither (a, b) nor $[a, b]$ can be negative.
33. Show why $ab/(a, b)$ must be an integer whenever (a, b) is defined.
34. Show that a and $-a$ have the same divisors and the same multiples.
35. Show that if (a, b) is defined, then it equals $(|a|, |b|)$, and that if $[a, b]$ is defined, then it equals $[|a|, |b|]$.
36. Prove that if $a \mid b$, then there exists an integer c such that $b = ac$.
37. Prove that if a and b are not both 0 and if there exists an integer c such that $b = ac$, then $a \mid b$.

1.2   The Division Algorithm

Turning Tricks into Methods

Our proof of the last theorem involved an idea that will be useful in the future. Let us try to write it down in a general form, as a theorem that may simply be quoted if the same trick is needed again. (Someone has said that in mathematics if a "trick" is used twice it becomes a "method.")

The idea occurred near the end of the proof when we expressed m as a multiple of $[a, b]$ plus an integer r, where $0 \le r < [a, b]$. (In the proof of the theorem the assumption that $[a, b]$ didn't divide m implied that $r > 0$.) This is really a familiar arithmetic fact. For example, one knows without doing the actual computation that if 592 is divided by 7 the answer is some integer q plus $r/7$, where

$r = 0, 1, 2, 3, 4, 5,$ or 6. Thus (multiplying by 7)

$$592 = 7q + r, \quad 0 \le r < 7.$$

In fact, if you and I both do this computation, we should expect to get the same values for q and r. In other words, the **quotient** q and **remainder** r are uniquely determined.

THE DIVISION ALGORITHM   Suppose a and b are integers, $a > 0$. Then there exist unique integers q and r, $0 \le r < a$, such that $b = aq + r$.

Proof   First we show q and r exist. Suppose the real number b/a is $q + \varepsilon$, where q is an integer and $0 \le \varepsilon < 1$. Then

$$b = a(q + \varepsilon) = aq + a\varepsilon.$$

From $b = aq + a\varepsilon$ it follows that $a\varepsilon$ is an integer, and $0 \le a\varepsilon < a$. (Why?) Set $a\varepsilon = r$.

Next we show q and r are unique. Suppose

$$b = aq + r, \quad 0 \le r < a,$$

$$b = aq' + r', \quad 0 \le r' < a.$$

Subtracting the two equations and transposing $r - r'$ gives

$$r' - r = a(q - q').$$

But adding the inequalities $-a < -r \le 0$ and $0 \le r' < a$ and dividing by a yields

$$-1 < \frac{r' - r}{a} < 1.$$

Since $(r' - r)/a$ is the integer $q - q'$ we see $q - q' = 0$. This implies $q = q'$ and $r = r'$.

Examples   If $a = 7$ and $b = 592$, then $q = 84$ and $r = 4$, since

$$592 = 7 \cdot 84 + 4.$$

If $a = 5$ and $b = 115$, then $q = 23$ and $r = 0$, since

$$115 = 5 \cdot 23 + 0.$$

If $a = 9$ and $b = 4$, then $q = 0$ and $r = 4$, since

$$4 = 9 \cdot 0 + 4.$$

If $a = 6$ and $b = -22$, then $q = -4$ and $r = 2$, since

$$-22 = 6(-4) + 2.$$

The Division Algorithm with a Calculator

If the numbers a and b are large it may save time to use a calculator to compute the q and r of the Division Algorithm. As we saw in its proof, q is just the largest integer less than or equal to b/a. If $a = 375$ and $b = 49162$, one calculator gives $b/a = 131.0986667$, and so $q = 131$.

Now $r = a(b/a - q)$. Thus in our example $r = 375(131.098667 - 131) = 37.0000125$ on the same calculator. Of course r is the integer 37, and the .0000125 is due to round-off error.

In general to find q and r we:

1. compute b/a; q is the largest integer not exceeding this number,
2. subtract q from b/a,
3. multiply by a to get r.

Looking for a Pattern

Number theory has the advantage that we may sometimes find theorems simply by experimentation. We will try to find some nonobvious relation or formula involving (a, b) or $[a, b]$. Let us see what happens when we combine (a, b) and $[a, b]$ in various ways. To keep things simple we will set $a = 6$ and let b vary.

a	b	(a, b)	$[a, b]$	$(a, b) + [a, b]$	$[a, b] - (a, b)$	$(a, b)[a, b]$
6	1	1	6	7	5	6
6	2	2	6	8	4	12
6	3	3	6	9	3	18
6	4	2	12	14	10	24
6	5	1	30	31	29	30
6	6	6	6	12	0	36
6	7	1	42	43	41	42

The only column showing much regularity is the one giving $(a, b)[a, b]$. Perhaps we should try a different value of a, say $a = 4$.

a	b	(a, b)	$[a, b]$	$(a, b)[a, b]$
4	1	1	4	4
4	2	2	4	8
4	3	1	12	12
4	4	4	4	16
4	5	1	20	20
4	6	2	12	24

It appears that perhaps $(a, b)[a, b] = ab$ for all integers a and b. We must be careful however; what if a or b were negative? It is easy to see from the definitions that (a, b) and $[a, b]$ are always positive, but ab might not be. We will avoid this problem by assuming that a and b are both positive.

THEOREM   If a and b are positive integers, then $(a, b)[a, b] = ab$.

Proof   In this proof we will leave out the reasons for several steps in the argument, asking the reader to supply them. The correct reasons will be found at the end of the problems for this section.

The equation we are trying to prove may be rewritten as

$$[a, b] = \frac{ab}{(a, b)}.$$

One way to prove this would be to show that the expression on the right satisfies the definition of $[a, b]$. It is easy to see that it is positive. [**Question 1: Why?**] Also $ab/(a, b)$ is an integer since it is a times $b/(a, b)$, and the latter is an integer. [**Question 2: Why?**]

Is $ab/(a, b)$ a common multiple of a and b? We have

$$\frac{ab/(a, b)}{a} = \frac{b}{(a, b)},$$

which we have just seen is an integer, so $ab/(a, b)$ is a multiple of a. In the same way we could prove that $ab/(a, b)$ is a multiple of b. Thus $ab/(a, b)$ is a common multiple of a and b, from which we conclude that

$$\frac{ab}{(a, b)} \geq [a, b]. \quad [\textbf{Question 3: Why?}]$$

If we could prove this inequality in the opposite direction we would be done. Let us start over by considering $ab/[a, b]$. This is an integer. [**Question 4: Why?**] Also

$$\frac{a}{ab/[a, b]} = \frac{[a, b]}{b},$$

which is an integer. [**Question 5: Why?**] Thus $ab/[a, b]$ divides a. A similar proof shows that $ab/[a, b]$ divides b. We see that $ab/[a, b]$ is a common divisor of a and b. Thus

$$\frac{ab}{[a, b]} \le (a, b). \quad [\textbf{Question 6: Why?}]$$

Since the two inequalities we have proved are equivalent to

$$ab \ge (a, b)[a, b] \quad \text{and} \quad ab \le (a, b)[a, b],$$

the theorem follows.

PROBLEMS FOR SECTION 1.2

A *Find the q and r guaranteed by the Division Algorithm for each pair a, b in problems 1 through 12.*

1. $a = 13, b = 380$
2. $a = 15, b = 421$
3. $a = 720, b = 155$
4. $a = 339, b = 17$
5. $a = 17, b = 51$
6. $a = 21, b = 105$
7. $a = 19, b = 0$
8. $a = 35, b = 0$
9. $a = 7, b = -30$
10. $a = 9, b = -29$
11. $a = 43, b = -500$
12. $a = 47, b = -500$

13. What are all the common divisors of 12 and 18?
14. What are all the common divisors of 45 and 75?
15. What are all the common multiples of 4 and 6?
16. What are all the common multiples of 27 and 18?

True-False. In the next eight problems tell which statements are true and give counterexamples for those that are false. Assume a, b, c, and d are arbitrary integers with a > 0 and c and d nonzero.

17. There exist integers q and r, $0 \le r < c$, such that $b = cq + r$.
18. There exist integers q and r, $0 \le r < |c|$, such that $b = cq + r$.
19. There exist integers q and r, $|r| \le a/2$, such that $b = aq + r$.

20. There exist integers q and r, $|r| < a/2$, such that $b = aq + r$.
21. The set of common divisors of b and c is the set of divisors of (b, c).
22. The set of common multiples of c and d is the set of multiples of $[c, d]$.
23. If b is a divisor of c, and $b > (c, d)$, then b is not a divisor of d.
24. If b is a multiple of c, and $b < [c, d]$, then b is not a multiple of d.

25. Prove that if $ab \neq 0$, then $(a, b)[a, b] = |ab|$.
26. Prove that $(a, a + 2)$ is 2 if a is even and 1 if a is odd.
27. Prove that if $a > 0$ then $[a, a + 2] = a(a + 2)/2$ if a is even and $a(a + 2)$ if a is odd.
28. Prove that if $x > 0$ then $(a, a + x)$ is (a, x).
29. Prove that if $d|a$, $d|b$, and $d|c$, and if x, y, and z are any integers, then d divides $ax + by + cz$.
30. Prove that if $a|b$, $b|c$, and $c|d$, then $a|d$.
31. Prove that if $a|b$ and $b|c$, then a divides $ax + by + cz$ for any integers x, y, and z.
32. Prove that if $b \neq 0$ and $a = bx + cy$, then $(b, c) \leq (a, b)$.
33. Prove that with the hypotheses of the last problem, $(b, c)|(a, b)$.
34. Show that if a and b are nonzero, then $(a, b) = (a + b, [a, b])$.

Answers to questions in the proof of the last theorem of this section.

1. By hypothesis $ab > 0$, and $(a, b) > 0$ by definition.
2. By the definition of (a, b).
3. Since $[a, b]$ is the *least* common multiple of a and b.
4. By the theorem at the end of Section 1.1.
5. By the definition of $[a, b]$.
6. Since (a, b) is the *greatest* common divisor of a and b.

1.3   The Euclidean Algorithm

Calculating the Greatest Common Divisor

Although we have proved various theorems about the greatest common divisor (a, b), we have no easy way to compute it. Of course we could simply list all the divisors of the smaller of a and b, say a, and look for the largest of these that also divides b. If a is large, however, this could be tedious. It turns out that the Division Algorithm yields a way to make the task easier.

Let us denote (a, b) by d, and suppose $0 < a < b$. We write

$$b = aq + r, \quad 0 \leq r < a.$$

Then $r = b - aq$, and since d divides both a and b, Theorem 1.2 implies that d divides r also. Since $r < a$ this fact may make finding d easier.

In fact since d divides both r and a, $d \leq (r, a)$ by the definition of the latter. But applying Theorem 1.2 to $b = aq + r$ shows, in the same way, that (a, r) divides b. Since (a, r) divides a by definition, we have $(a, r) \leq (a, b) = d$ by the definition of (a, b).

We see from the above paragraph that $(a, b) = (r, a)$. Actually nothing more than the equation $b = aq + r$ was used in the proof, and the fact that q and r came from the Division Algorithm was irrelevant. We have proved the following:

LEMMA   If $a = bk + c$, then $(a, b) = (b, c)$.

This fact is especially useful when applied to the equation of the Division Algorithm as above because $a > r$ and $b > a$, so we have replaced a and b by a smaller pair. In fact the process can be repeated until the gcd is obvious.

Example   Suppose we want $(504, 123)$. Set $a = 123$ and $b = 504$. Now

$$504 = 123 \cdot 4 + 12.$$

Thus $r = 12$ and $(123, 504) = (12, 123)$. Likewise

$$123 = 12 \cdot 10 + 3,$$

so $(12, 123) = (3, 12)$.

Although clearly $(3, 12) = 3$, let us see how far we can carry this process. Now

$$12 = 3 \cdot 4 + 0,$$

and so $(3, 12) = (0, 3)$. This is clearly 3 since any positive number divides 0. Thus $(504, 123) = 3$.

THE EUCLIDEAN ALGORITHM   Suppose a and b are integers, $a > 0$. Apply the Division Algorithm repeatedly as follows:

$$b = aq_1 + r_1, \quad 0 \leq r_1 < a,$$

$$a = r_1 q_2 + r_2, \quad 0 \leq r_2 < r_1,$$

$$r_1 = r_2 q_3 + r_3, \quad 0 \leq r_3 < r_2,$$

$$\vdots$$

$$r_{n-2} = r_{n-1} q_n + r_n, \quad 0 < r_n < r_{n-1},$$

$$r_{n-1} = r_n q_{n+1}.$$

Suppose r_n is the last nonzero remainder. Then $(a, b) = r_n$.

Proof   The process must terminate, since the remainders form a strictly decreasing sequence of nonnegative integers. Then, as we have seen,

$$(a, b) = (a, r_1) = (r_1, r_2) = \cdots = (r_{n-1}, r_n) = (r_n, 0) = r_n.$$

Example   We will compute $(158, 188)$, using arrows to emphasize the pattern.

$$188 = 158 \cdot 1 + 30$$
$$158 = 30 \cdot 5 + 8$$
$$30 = 8 \cdot 3 + 6$$
$$8 = 6 \cdot 1 + 2$$
$$6 = 2 \cdot 3 + 0$$

Thus $(158, 188) = 2$.

Solving the Euclidean Algorithm Equations Backwards

It is possible to use the Euclidean Algorithm equations to write (a, b) in the form $ax + by$. We start with the equation having (a, b) as a remainder and work backwards. We will do this for the above example.

$$2 = 8 - 6 \cdot 1$$
$$= 8 - (30 - 8 \cdot 3)1 \qquad\qquad = -30 + 8 \cdot 4$$
$$= -30 + (158 - 30 \cdot 5)4 \quad = 158 \cdot 4 - 30 \cdot 21$$
$$= 158 \cdot 4 - (188 - 158)21 = -188 \cdot 21 + 158 \cdot 25$$

Thus $(158, 188) = 158x + 188y$ for $x = 25$ and $y = -21$.

If either a or b is negative we simply deal with $|a|$ and $|b|$ and insert the correct signs at the end of the problem. For example, suppose we want to find x and y such that $ax + by = (a, b)$ for $a = -56$ and $b = -21$. We work with 21 and 56.

$$56 = 21 \cdot 2 + 14$$

$$21 = 14 \cdot 1 + 7$$

$$14 = 7 \cdot 2 + 0$$

Thus $7 = (21, 56) = (-56, -21)$. Also

$$7 = 21 - 14 \cdot 1$$
$$= 21 - (56 - 21 \cdot 2)1 = -56 + 21 \cdot 3$$
$$= (-56)(1) + (-21)(-3).$$

We see $x = 1$ and $y = -3$.

The proof that this can always be done will be left for the problems at the end of this section.

THEOREM   If (a, b) is defined then there exist integers x and y such that $(a, b) = ax + by$.

DEFINITION   linear combination

We say k is a **linear combination** of a and b in case there exist integers x and y such that $k = ax + by$.

The Duality of the gcd and lcm

The reader may have noticed a certain duality between the concepts of least common multiple and greatest common divisor. Replacing "least" with "greatest" and "multiple" with "divisor" changes the definition of the former into that of the latter. New (possible) theorems can be created in a like way. For example, the following theorem restates the last theorem of Section 1.1 as its part (a); part (b) is the dual result.

THEOREM 1.3

    a. If m is any common multiple of a and b, then $[a, b]$ divides m.

    b. If d is any common divisor of a and b, not both of which are 0, then d divides (a, b).

Proof of (b)   By the last theorem there exist x and y such that $(a, b) = ax + by$, which is divisible by d by Theorem 1.2.

The Euclidean Algorithm on a Computer

The Euclidean Algorithm is one of those repetitive computations that computers do so well. Here is a BASIC program computing (a, b). It assumes that $a > 0$.

```
10  INPUT A,B
20  LET Q = INT(B/A)
30  LET R = B - A*Q
40  IF R = 0 GOTO 80
50  LET B = A
60  LET A = R
70  GOTO 20
80  PRINT A
90  END
```

PROBLEMS FOR SECTION 1.3

 Find (a, b) for each pair given in problems 1 through 8 using the Euclidean Algorithm. Then solve the equations backwards to find x and y such that $ax + by = (a, b)$.

1. $a = 217, b = 341$
2. $a = 117, b = 247$
3. $a = 143, b = 451$
4. $a = 165, b = 465$
5. $a = 89, b = 55$
6. $a = 123, b = 76$
7. $a = -899, b = 2030$
8. $a = -4050, b = -1728$

9. Find x and y such that $26x + 14y = (26, 14)$ with x positive but as small as possible.
10. Find x and y such that $27x + 15y = (27, 15)$ with x positive but as small as possible.
11. Use problem 1 to solve $217x + 341y = 62$.
12. Use problem 2 to solve $117x + 247y = 39$.
13. Find all solutions to $2x + 3y = 50$ in positive integers.
14. Find all solutions to $3x + 4y = 60$ in positive integers.
15. Find all solutions to $4x + 6y = 60$ in positive integers.
16. Find all solutions to $6x + 9y = 91$ in positive integers.

 17. Why is $4x + 6y = 25$ unsolvable?
18. Why is $361x + 2109y = 1000$ unsolvable?
19. Determine all integers x such that $2x + 3y = 1$ is solvable.
20. Determine all integers x such that $3x + 2y = 4$ is solvable.

 21. Prove that if v is a linear combination of w and x, and if w and x are each linear combinations of y and z, then v is a linear combination of y and z.
22. Use the previous problem to prove the last theorem of this section.
23. Let S be a set of integers containing at least one nonzero element and having the property that whenever a and b are in S, then so are $a + b$ and $a - b$. Show that S contains positive elements.

24. Let c be the least positive element of the set S of the previous problem. Show that S is exactly all multiples of c. (**Hint:** Use the Division Algorithm.)
25. Use the previous problem to provide another proof of the last theorem of this section. (**Hint:** Let S be all linear combinations of a and b.)
26. Use the last theorem of this section to prove Theorem 1.4.
27. Write a computer program that calculates not only (a, b) but also x and y such that $ax + by = c$. (**Hint:** Instead of keeping track of the q's and r's so as to solve the equations backwards, write r_i as a linear combination of a and b when it is computed.)
28. Why is the assumption that a and b are not both 0 needed in part (b) of Theorem 1.3 but not in part (a)?

1.4   Linear Combinations

When $ax + by = c$ Is Solvable

We can find x and y such that $ax + by = c$ not only when $c = (a, b)$, but also whenever c is any multiple of (a, b). For if $c = k(a, b)$ and $ax + by = (a, b)$, then $a(kx) + b(ky) = c$. Conversely, if $ax + by = c$, then (a, b) must divide c by Theorem 1.2.

THEOREM 1.4   Given integers a, b, and c with a and b not both 0, there exist x and y such that $ax + by = c$ if and only if $(a, b)|c$.

Corollary   There exist integers x and y such that $ax + by = 1$ if and only if $(a, b) = 1$.

Corollary   If $(a, b) = 1$ and $(a', b) = 1$, then $(aa', b) = 1$.

Proof   By the preceding corollary there exist integers x, y, x', and y' such that

$$ax + by = 1 \quad \text{and} \quad a'x' + by' = 1.$$

Multiplying these equations gives

$$(aa')(xx') + b(axy' + a'x'y + byy') = 1.$$

Thus (again by the first corollary) $(aa', b) = 1$.

DEFINITION   relatively prime
    We say a and b are **relatively prime** in case $(a, b) = 1$. In a certain sense if a and b are relatively prime, then a and b "have nothing to do with each other." The next theorem illustrates this idea.

THEOREM 1.5 If $(a, b) = 1$ and if $a|bc$, then $a|c$.

Proof By the first corollary there exist x and y such that $ax + by = 1$. Then $acx + bcy = c$. Since $a|a$ and $a|bc$ we apply Theorem 1.2 to conclude $a|c$.

WARNING: One of the most common mistakes made by beginning number theory students is to assume that if $a|bc$, then $a|b$ or $a|c$. This is not generally true, and the reader should construct a counterexample before proceeding.

If we remove the "common part" (a, b) from a and b, we would expect the remaining integers to have nothing left in common. The following theorem expresses this idea precisely.

THEOREM 1.6 If $(a, b) = d$, then a/d and b/d are relatively prime.

Proof We can write $ax + by = d$. Then $(a/d)x + (b/d)y = 1$, and so a/d and b/d are relatively prime by the first Corollary.

All Solutions to ax + by = c

If $(a, b)|c$, our Euclidean Algorithm method enables us to find one solution to $ax + by = c$. Consider the equation $9x + 24y = 15$, for example. We write

$$24 = 9 \cdot 2 + 6$$

$$9 = 6 \cdot 1 + 3$$

$$6 = 3 \cdot 2 + 0.$$

Since $(9, 24) = 3$ and $3|15$ we know a solution exists. In fact

$$3 = 9 - 6 \cdot 1$$

$$= 9 - (24 - 9 \cdot 2)1 = 9 \cdot 3 - 24 \cdot 1.$$

Then $15 = 3 \cdot 5 = 9 \cdot 15 - 24 \cdot 5$. We see $x = 15$ and $y = -5$ is a solution. But there are others, for example $x = -1$ and $y = 1$.

We will study how to find all solutions to $ax + by = c$, given a particular one, say x_0, y_0. Suppose

$$ax_0 + by_0 = c \qquad \text{and} \qquad ax + by = c;$$

so x, y is another solution. Then

$$a(x - x_0) = b(y_0 - y).$$

We see that $a \mid b(y_0 - y)$. If we knew that a and b were relatively prime, then we could conclude by Theorem 1.5 that a divided $y_0 - y$; but of course it is possible that $(a, b) = d > 1$.

We will get around this by using the equation

$$\frac{a}{d}(x - x_0) = \frac{b}{d}(y_0 - y),$$

where a/d and b/d are relatively prime by Theorem 1.6. The equation implies $a/d \mid y_0 - y$.

Suppose $y_0 - y = (a/d)t$, where t is some integer. Then our equation yields

$$x - x_0 = \frac{b}{a}(y_0 - y) = \frac{b}{d} \cdot t.$$

Solving for x and y gives

$$x = x_0 + \frac{b}{d} \cdot t \quad \text{and} \quad y = y_0 - \frac{a}{d} \cdot t.$$

Let us review what we have done so far. We assumed x_0, y_0 was a particular solution to $ax + by = c$, and also that x, y was some other solution. This forced us to the conclusion that

$$\left.\begin{array}{l} x = x_0 + \dfrac{b}{d} \cdot t \\[2mm] y = y_0 - \dfrac{a}{d} \cdot t, \end{array}\right\} \tag{$*$}$$

where t is some integer and $d = (a, b)$.

We see any other solution must be of the form $(*)$. The question remains whether any pair x, y of the form $(*)$ must be a solution to the original equation. Plugging in such a pair shows that this is the case. Indeed

$$a\left(x_0 + \frac{b}{d} \cdot t\right) + b\left(y_0 - \frac{a}{d} \cdot t\right) = ax_0 + by_0 + \frac{abt}{d} - \frac{abt}{d} = c.$$

THEOREM 1.7 Let x_0, y_0 be a particular solution to $ax + by = c$. Then all solutions are given by

$$x = x_0 + \frac{b}{d} \cdot t, \quad y = y_0 - \frac{a}{d} \cdot t,$$

as t runs through the integers, where $d = (a, b)$.

Example We have seen that one solution to $9x + 24y = 15$ is $x_0 = 15$, $y_0 = -5$, and that $(9, 24) = 3$. Then all solutions are

$$x = 15 + \frac{24}{3}t = 15 + 8t, \qquad y = -5 - \frac{9}{3}t = -5 - 3t.$$

Taking $t = -2$ gives the solution $x = -1$, $y = 1$ noted previously.

Example Find all positive integers x, y such that $4x + 6y = 100$.

We note that $(4, 6) = 2$ and $2 \mid 100$, so solutions do exist. In fact $x = 25$ and $y = 0$ is one solution, although not in positive integers.

The general solution is

$$x = 25 + \frac{6}{2}t = 25 + 3t, \qquad y = 0 - \frac{4}{2}t = -2t.$$

In this case we want $x = 25 + 3t > 0$ and $y = -2t > 0$. The first inequality amounts to $25 > -3t$ or $t > -8\frac{1}{3}$, while $-2t > 0$ says $t < 0$. Thus t runs from -1 to -8, and we get the solutions shown below.

t	-1	-2	-3	-4	-5	-6	-7	-8
x	22	19	16	13	10	7	4	1
y	2	4	6	8	10	12	14	16

Example Find all solutions to $15x + 16y = -1000$ with $x \le 0$ and $y \le 0$.

Clearly $(15, 16) = 1$ and

$$15(-1) + 16(1) = 1.$$

Thus $x_0 = 1000$ and $y_0 = -1000$ is one solution to the original equation. The general solution is

$$x = x_0 + bt = 1000 + 16t, \qquad y = y_0 - at = -1000 - 15t.$$

Also $x = 1000 + 16t \le 0$ says

$$16t \le -1000 \qquad \text{or} \qquad t \le -62\frac{1}{2},$$

and $y = -1000 - 15t \le 0$ says

$$15t \ge -1000 \qquad \text{or} \qquad t \ge -66\frac{2}{3}.$$

Since t is an integer we see $t = -63$, -64, -65, or -66.

$-66\frac{2}{3}$ ... -66 ... -65 ... -64 ... -63 ... $-62\frac{1}{2}$ ... -62

t	-63	-64	-65	-66
x	-8	-24	-40	-56
y	-55	-40	-25	-10

Example   A man paid \$11.37 for some 39-cent pens and 69-cent pens. How many of each did he buy?

Suppose he bought x pens at 39 cents and y pens at 69 cents. Then we wish to solve

$$39x + 69y = 1137, \quad x \geq 0, \, y \geq 0.$$

We have $(39, 69) = 3$ and $3 | 1137$, so the equation has solutions (although we still don't know if any are nonnegative). Since we found $(39, 69) = 3$ without recourse to the Euclidean Algorithm we will divide by 3 and solve the simpler equivalent system

$$13x + 23y = 379, \quad x \geq 0, \, y \geq 0.$$

Using the Euclidean Algorithm on 13 and 23 and solving the equations backwards leads to the equation

$$13(-7) + 23(4) = 1.$$

Thus we have the particular solution $x_0 = 379(-7) = -2653$, $y_0 = 379(4) = 1516$. The general solution is

$$x = -2653 + 23t, \qquad y = 1516 - 13t.$$

Here $x \geq 0$ and $y \geq 0$ amount to

$$t \geq \frac{2653}{23} = 115\frac{8}{23} \quad \text{and} \quad t \leq \frac{1516}{13} = 116\frac{8}{13}.$$

Thus $t = 116$, $x = -2653 + 23(116) = 15$, and $y = 1516 - 13(116) = 8$. The man bought 15 pens at 39 cents and 8 pens at 69 cents.

PROBLEMS FOR SECTION 1.4

A *In problems 1 through 6 tell whether or not the equation has a solution.*

1. $3x + 5y = 50001$
2. $6x + 9y = 60001$
3. $21x - 14y = 10000$
4. $-12x + 42y = 366$
5. $529x + 2024y = 391$
6. $851x + 1147y = 481$

Use the Euclidean Algorithm method to find one solution to the equation in problems 7 through 12.

7. $7x + 20y = 3$ 8. $8x + 21y = 5$
9. $66x + 51y = 300$ 10. $65x + 50y = 300$
11. $200x - 300y = 400$ 12. $55x + 200y = -100$

In problems 13 through 19 find all solutions with x and y positive.

13. $5x + 6y = 100$ 14. $6x + 7y = 200$
15. $6x + 8y = 120$ 16. $9x + 6y = 150$
17. $121x + 561y = 13200$ 18. $169x + 663y = 2340$
19. $621x + 1026y = 49194$

20. If $abc \neq 0$, is it possible for $ax + by = c$ to have infinitely many solutions in positive integers?
21. For what triples a, b, c is it true that for each integer x there is an integer y such that $ax + by = c$?
22. A girl spent $100.64 on records. Some cost $4.98 and some $5.98. How many of each did she buy?
23. A man bought three dozen oranges and two dozen apples. His change from a $10 bill was $1.96. One orange costs more than 10 cents and one apple more than 15 cents. How much does an orange cost?
24. A roadside stand bought 11 large baskets of eggs from a farmer and sold 39 small baskets of eggs. There were 19 eggs left over. How many eggs does a large basket hold?
25. Farmer Jones owes Farmer Brown $10. Neither has any cash, but Jones has 14 cows which he values at $185 each. He suggests paying his debt in cows, with Brown making change by giving Jones some of his pigs at $110 each. Is this possible, and how?
26. Brown wants his $10, but says his pigs are worth $111 each. Is the deal still possible, and how?
27. Jones offers to trade his cows for $184 each if Brown accepts $110 per pig. Now what?
28. Harry bought some 39-cent pens. He paid with a $10 bill, getting 42 cents in coins in change (plus maybe some bills). How many did he buy?
29. The same Harry as in the last problem bought 28 more-expensive pens at another store, receiving 44 cents change in coins from a $20 bill, plus perhaps some bills. How much were the pens?
30. A retailer would like to sell big boxes of expensive Christmas cards for less than $15. He wants to get 34 cents for each card, and wants the price of a box to end in 99—for example, $7.99. If this is impossible he would like it to end in 98, 97, or 96, etc., in order of decreasing desirability. How many cards should he put in a box?
31. Alan paid $133.98 for some $1.29 records and $2.49 books. He didn't buy 100 records and 2 books, even though this adds up to the right amount. How many of each *did* he buy?

32. A woman paid a $9.75 debt using only dimes and quarters. In how many ways is this possible?

33. Tina spent $86.66 on some $1.47 puzzles and on 371 erasers. How much does an eraser cost?

34. Prove that if $a|bc$, then $a|(a, b)c$.

35. Give a proof of Theorem 1.6 using nothing more than the definition of (a, b).

36. Show that if $(a, b)|c$ and if $ab < 0$, then $ax + by = c$ has at least one solution with x and y positive.

1.5 Congruences

A Table for (a, b)

At the beginning of this chapter we discovered a relation between (a, b) and $[a, b]$ by looking at a table. We will return to this technique, fixing a at 6 and letting b vary.

a	6	6	6	6	6	6	6	6	6	6	6	6	6
b	1	2	3	4	5	6	7	8	9	10	11	12	13
(a, b)	1	2	3	2	1	6	1	2	3	2	1	6	1

The values of (a, b) seem to repeat the pattern $1-2-3-2-1-6$ again and again. It appears that whatever $(6, b)$ is, $(6, b + 6)$ will be the same, as will $(6, b + 12)$, $(6, b + 18)$, etc. For example $(6, 1) = (6, 7) = (6, 13) = 1$. The numbers 1, 7, and 13 have the property that any two of them differ by a multiple of 6. This idea is important enough to have its own symbolism.

DEFINITION   congruence

Let $m > 0$. We say that a is **congruent to** b **modulo** m and write

$$a \equiv b \quad (\text{mod } m)$$

in case $m|b - a$.

Examples

Since $6|6 = 7 - 1$, $1 \equiv 7$ (mod 6).

Since $6|12 = 13 - 1$, $1 \equiv 13$ (mod 6).

Since $6 \nmid 5 = 7 - 2$, $2 \not\equiv 7$ (mod 6).

Why Congruence?

The reader may think that we do not need a new and slightly longer way to say $m \mid b - a$. However the notation $a \equiv b \pmod{m}$ is useful because it suggests an equation, and in fact congruences share many properties with equations. A symbolism that emphasizes this similarity is useful for suggesting new results and making them easier to remember.

We will soon see for example that if $a \equiv b \pmod{m}$ and $a' \equiv b' \pmod{m}$, then $aa' \equiv bb' \pmod{m}$. This is easy to assimilate as analogous to "multiplying" equations. Without the new symbolism, however, the result would say that if $m \mid b - a$ and $m \mid b' - a'$, then $m \mid bb' - aa'$, a much more awkward formulation.

The next theorem lists a number of ways in which congruences behave like equations. Readers are encouraged to try to find their own proofs before reading those given. Part (e) is the hardest.

THEOREM 1.8   Let $m > 0$. All congruences in parts (a) through (e) in this theorem are modulo m, but the "(mod m)" has been omitted. Let a, b, c, d, a', and b' be integers, with $d > 0$.

a. $a \equiv a$.
b. If $a \equiv b$,   then $b \equiv a$.
c. If $a \equiv b$ and $b \equiv c$,   then $a \equiv c$.
d. If $a \equiv b$ and $a' \equiv b'$,   then $a + a' \equiv b + b'$.
e. If $a \equiv b$ and $a' \equiv b'$,   then $aa' \equiv bb'$.
f. If $a \equiv b$ (mod m) and $d \mid m$,   then $a \equiv b$ (mod d).

Proof

a. Always $m \mid a - a = 0$.
b. If $b - a = mk$, then $a - b = m(-k)$.
c. We have $m \mid b - a$ and $m \mid c - b$, so $m \mid c - a = (c - b) + (b - a)$ by Theorem 1.2.
d. Since $m \mid b - a$ and $m \mid b' - a'$, we have $m \mid (b - a) + (b' - a')$ $= (b + b') - (a + a')$ by Theorem 1.2.
e. Since $m \mid b - a$ and $m \mid b' - a'$, we have $m \mid (b - a)b' + a(b' - a')$ $= bb' - aa'$ by Theorem 1.2.
f. Since $m \mid b - a$ and $d \mid m$, $d \mid b - a$ by Theorem 1.1.

Note: The proof for part (e) uses the common mathematical trick of adding and subtracting the same thing to get an agreeable form, since

$$bb' - aa' = bb' - ab' + ab' - aa' = (b - a)b' + (b' - a')a.$$

DEFINITION   least residue
If $m > 0$ and r is the remainder when the Division Algorithm is used to divide b by m, then r is called the **least residue of b modulo m**.

Examples   The least residue of 12 modulo 7 is 5. The least residue of 7 modulo 12 is 7. The least residue of 20 modulo 4 is 0. The least residue of -12 modulo 7 is 2. The least residue of -3 modulo 7 is 4.

THEOREM 1.9   Let $m > 0$.

   a. Two integers are congruent modulo m if and only if they have the same least residue modulo m.
   b. If $r \equiv b \pmod{m}$ and $0 \le r < m$, then r is the least residue of b modulo m.

Proof

   a. Let the two numbers be b and b'. Suppose

   $$b = mq + r, \quad 0 \le r < m, \qquad b' = mq' + r', \quad 0 \le r' < m.$$

If $b \equiv b' \pmod{m}$, then

   $$m \mid (b' - b) - m(q' - q) = r' - r.$$

   But (exactly as in the proof of the uniqueness of the Division Algorithm) $-m < r' - r < m$. We conclude that $r - r' = 0$.
      Conversely if $r = r'$ then $b' - b = m(q' - q)$ and so $m \mid b' - b$.
   b. Let $b - r = mq$. Then $b = mq + r$ and r is the remainder in the Division Algorithm by the uniqueness of that remainder.

Example   Find the least residue (mod 31) of $33 \cdot 26^2$.
   We have $33 \equiv 2 \pmod{31}$ and $26 \equiv -5 \pmod{31}$. By part (e) of Theorem 1.8 we conclude that

   $$33 \cdot 26^2 \equiv 2(-5)^2 = 50 \equiv 19 \pmod{31}.$$

Since $0 \le 19 < 31$, by Theorem 1.9 the least residue is 19.

   The congruence properties of Theorem 1.8 allow the substitution of congruent values in a polynomial. Suppose the least residue of $3(53) + 27^2 \pmod{7}$ is desired. Since $53 \equiv 4 \pmod{7}$ and $27 \equiv -1 \pmod{7}$, we have

   $$3(53) + 27^2 \equiv 3(4) + (-1)^2 = 13 \equiv 6 \pmod{7}, \tag{$*$}$$

and the least residue is 6. The justification for the first congruence of $(*)$ is as follows, where the modulus is 7:

   $$53 \equiv 4,$$

   $$3(53) \equiv 3(4) \quad \text{(part (e))},$$

   $$27 \equiv -1,$$

$$27^2 = 27 \cdot 27 \equiv (-1)(-1) = (-1)^2 \quad \text{(part (e))},$$

$$3(53) + 27^2 \equiv 3(4) + (-1)^2 \quad \text{(part (d))}.$$

Although both sides of a congruence may be multiplied by the same number by part (e) of Theorem 1.8, division is another matter. For example

$$2 \cdot 1 \equiv 2 \cdot 3 \quad \text{(mod 4)},$$

but $1 \not\equiv 3$ (mod 4). The following theorem tells when division is allowed.

THE CANCELLATION THEOREM If $(a, b) = 1$, then $ax \equiv ax'$ (mod b) implies $x \equiv x'$ (mod b).

This theorem is a special case of the following more general theorem.

THEOREM 1.10 If $(a, b) = d$, then

$$ax \equiv ax' \quad \text{(mod } b) \quad \text{if and only if} \quad x \equiv x' \quad \text{(mod } b/d).$$

Proof If $ax \equiv ax'$ (mod b), then b divides $a(x' - x)$. Say $a(x' - x) = bk$. Then $(a/d)(x' - x) = (b/d)k$. Since $(a/d, b/d) = 1$ by Theorem 1.6, b/d divides $x' - x$ by Theorem 1.5.
   Conversely, if $x \equiv x'$ (mod b/d), then $x' - x = (b/d)k$ for some k. Then $ax' - ax = b(a/d)k$, which shows that $ax \equiv ax'$ (mod b).

A Congruence Property of (a, b)

We return to our observation of the periodicity of (a, b) as b runs through the integers, made in the beginning of this section.

THEOREM 1.11 If $b \equiv b'$ (mod a), then $(a, b) = (a, b')$.

Proof Let $b' - b = qa$. Then $b' = qa + b$. By the lemma in Section 1.3, $(b', a) = (a, b)$.

PROBLEMS FOR SECTION 1.5

A *In the first 18 problems find the least residue of b modulo m. Problems 9 through 14 are for students with calculators; see the method of Section 1.2. Theorem 1.8 may be useful in problems 15 through 18.*

1. $m = 7, b = 100$
2. $m = 8, b = 77$
3. $m = 50, b = 17$
4. $m = 51, b = 19$
5. $m = 50, b = -12$
6. $m = 51, b = -30$

7. $m = 7, b = -100$ 8. $m = 8, b = -77$

9. $m = 453, b = 613571$ 10. $m = 461, b = 916215$

11. $m = 61462, b = 818886$ 12. $m = 9162, b = 201563$

13. $m = 41522, b = -16115$ 14. $m = 91631, b = -2152$

15. $m = 63, b = 752 \cdot 571$ 16. $m = 51, b = 414 \cdot 566$

17. $m = 71, b = 72 \cdot 73 \cdot 74$ 18. $m = 82, b = 80 \cdot 81 \cdot 85$

19. Find the least residue of $5k$ modulo 7 for $k = 1, 2, 3, 4, 5, 6, 7$.
20. Find the least residue of $7k$ modulo 5 for $k = 1, 2, 3, 4, 5$.
21. What is the least residue of 100^6 modulo 49?
22. What is the least residue of 49^4 modulo 23?

B   23. Suppose $a \equiv 2$ (mod 17), $b \equiv 4$ (mod 17), and $c \equiv 5$ (mod 17). What is the least residue of $a + bc$ (mod 17)?
24. With the assumptions of problem 23, what is the least residue of $a^2 + b^2 + c^2$ (mod 17)?
25. What is the least residue of 50^{99} (mod 7)?
26. What is the least residue of 50^{99} (mod 17)?

In the next 3 problems use Theorem 1.10 to give a congruence equivalent to the one given with the numbers as small as possible.

27. $6x \equiv 14y$ (mod 10)   28. $6x \equiv 14y$ (mod 15)

29. $12x \equiv 30y$ (mod 15)

30. Let r and r' be the least residues of b and $-b$ modulo m. What is $r + r'$?

C   31. Give another proof of part (e) of Theorem 1.8 by first showing that if $a \equiv b$ (mod m), then $ac \equiv bc$ (mod m) for any c.
32. Show that if r is the least residue of b modulo m, then $r \equiv b$ (mod m).
33. Show that if $x \equiv y$ (mod m), then

$$ax^2 + bx + c \equiv ay^2 + by + c \pmod{m}$$

for any integers a, b, and c.
34. Show that if $a \equiv a'$ (mod m) and $b \equiv b'$ (mod m), then $a - b \equiv a' - b'$ (mod m).

1.6   Mathematical Induction

THEOREM 1.12   If $a \equiv b$ (mod m), then $a^n \equiv b^n$ (mod m) for any positive integer n.

Proof   By part (e) of Theorem 1.8 we can multiply the congruence $a \equiv b$ (mod m) by itself, getting

$$a^2 \equiv b^2 \pmod{m}.$$

Multiplying this by the original congruence yields

$$a^3 \equiv b^3 \pmod{m}.$$

Clearly this process may be repeated as many times as desired, until we have

$$a^n \equiv b^n \pmod{m}.$$

Analysis of the Proof

The proof given above is convincing because it is obvious that one can always take the next step, from $a \equiv b$ to $a^2 \equiv b^2$, from $a^2 \equiv b^2$ to $a^3 \equiv b^3$, and so on. Other statements true for all positive integers n, however, even though easy enough to confirm for any particular value of n, may not be so obviously true in general. Consider, for example, the following theorem.

THEOREM A   If n is any positive integer, then $4^n \equiv 3n + 1 \pmod{9}$.

If $n = 1$, this says

$$4 \equiv 3 + 1 \pmod{9},$$

which is true. Likewise for $n = 2$ we have

$$16 \equiv 6 + 1 \pmod{9},$$

and for $n = 3$ we have

$$64 \equiv 9 + 1 \pmod{9},$$

which are both also true. These particular cases do not convince us of the general truth of Theorem A, however. For this we need a more formal proof structure, that of Mathematical Induction.

Induction is a general method of proving facts to be true for each positive integer. The student usually is introduced to it as a means of proving formulas for sums, for example

$$1 + 2 + 3 + \cdots + n = \frac{n(n + 1)}{2}.$$

These sum formulas are often the only applications of induction the student sees, and he or she may form the opinion that its usefulness is quite limited. Nothing could be further from the truth. In this section we will give examples of induction proofs of various types. We start by formally stating the induction principle.

Principle of Mathematical Induction

Let $S(n)$ be a statement involving the positive integer n. Suppose

1. $S(1)$ is true.
2. If $S(k)$ is true for some integer k, then $S(k + 1)$ is also true.

Then $S(n)$ is true for all positive integers n.

We illustrate the principle by using it to prove Theorem 1.12 and Theorem A.

THEOREM 1.12   If $a \equiv b$ (mod m), then

$a^n \equiv b^n$ (mod m)   for all positive integers n.

Proof *(Induction)*   Let $S(n)$ be the statement

$a^n \equiv b^n$ (mod m).

1. The statement $S(1)$ says

$a \equiv b$ (mod m),   which is true by hypothesis.

2. Suppose $S(k)$ is true, that is

$a^k \equiv b^k$ (mod m).

Multiplying this by $a \equiv b$ (mod m) yields $a^{k+1} \equiv b^{k+1}$ (mod m), which is $S(k + 1)$.

Thus we have verified (1) and (2) of the induction principle. Then by that principle $a^n \equiv b^n$ (mod m) for all positive integers n.

THEOREM A   If n is any positive integer, then $4^n \equiv 3n + 1$ (mod 9).

Proof *(Induction)*   Let $S(n)$ be the statement

$4^n \equiv 3n + 1$ (mod 9).

1. The statement $S(1)$ is

$4 \equiv 3 + 1$ (mod 9),   which is true.

2. Assume $S(k)$, that is

$4^k \equiv 3k + 1$ (mod 9).

Clearly $4 \equiv 4$ (mod 9). Multiplying these congruences gives

$4^{k+1} \equiv 4(3k + 1)$ (mod 9).

The right side of this congruence is

$$(3 + 1)(3k + 1) = 9k + 3(k + 1) + 1.$$

Thus

$$4(3k + 1) \equiv 3(k + 1) + 1 \quad (\mathrm{mod}\ 9),$$

since subtracting the sides of this congruence gives $-9k$. By Theorem 1.8 part (c) we have

$$4^{k+1} \equiv 3(k + 1) + 1 \quad (\mathrm{mod}\ 9),$$

and this is $S(k + 1)$.

We have verified (1) and (2), and so by mathematical induction $4^n \equiv 3n + 1$ (mod 9) for all positive integers n.

Is Induction True?

Students often have trouble accepting the induction principle. The usual objection is that in part (2) we are "assuming what we are trying to prove." This is not the case, and the confusion arises because of inattention to the domain of the variable n. We are trying to prove that $S(n)$ holds for all positive integers n. In part (2) we assume merely that $S(k)$ holds for one particular integer k, and show $S(k + 1)$ also holds.

The induction principle is analogous to a string of dominoes, set up in a single line (see Figure 1-1). Suppose that

1. I knock down the first domino.
2. The dominoes are so arranged that if one falls down, then so does the next one.

Then every domino will fall down, no matter how many there are.

We will now give more examples of induction proofs. As is customary, the statement $S(n)$ will not be labeled explicitly as such.

Figure 1-1

THEOREM   If X is any real number greater than -1, then

$$(1 + X)^n \geq 1 + nX$$

for all positive integers n.

Proof *(Induction)*

   1. If $n = 1$ the inequality is

$$1 + X \geq 1 + X, \qquad \text{which is clearly true.}$$

   2. Suppose $(1 + X)^k \geq 1 + kX$. Since $1 + X > 0$ we can multiply through by $1 + X$ to get

$$
\begin{aligned}
(1 + X)^{k+1} &\geq (1 + X)(1 + kX) \\
&= 1 + (k + 1)X + kX^2 \geq 1 + (k + 1)X.
\end{aligned}
$$

This is the inequality for $n = k + 1$.

Thus by the induction principle $(1 + X)^n \geq 1 + nX$ for all positive integers n.

THEOREM 1.13   If $a_1, a_2, \ldots, a_n$ are relatively prime to b, then so is $a_1 a_2 \cdots a_n$.

Proof   The proof will be by induction on n.

   1. If $n = 1$, then we are given that $(a_1, b) = 1$, which is also what is to be proved.

   2. Now assume the theorem is true for some integer k, and that $a_1, a_2, \ldots, a_k, a_{k+1}$ are all relatively prime to b. By the induction hypothesis $(a_1 \cdots a_k, b) = 1$, and it is given that $(a_{k+1}, b) = 1$. Then by the second corollary after Theorem 1.4 we have $(a_1 \cdots a_k a_{k+1}, b) = 1$.

Thus by the induction principle the theorem holds for all positive integers n.

THE SUM OF AN ARITHMETIC PROGRESSION

$$A + (A + D) + (A + 2D) + \cdots + (A + (n-1)D) = \frac{n(2A + (n-1)D)}{2}$$

for all positive integers n.

Proof *(Induction)*

   1. If $n = 1$ the left side is A and the right side is $1(2A + 0 \cdot D)/2 = A$.

   2. Assume

$$A + (A + D) + \cdots + (A + (k-1)D) = \frac{k(2A + (k-1)D)}{2}.$$

Adding $A + kD$ to both sides gives

$$A + (A + D) + \cdots + (A + (k - 1)D) + (A + kD)$$

$$= \frac{k(2A + (k - 1)D)}{2} + (A + kD)$$

$$= \frac{2Ak + k(k - 1)D + 2A + 2kD}{2}$$

$$= \frac{2A(k + 1) + kD(k + 1)}{2} = \frac{(k + 1)(2A + kD)}{2}.$$

This is the formula for $n = k + 1$.

Thus by mathematical induction the formula is true for all positive integers n.

THE SUM OF A GEOMETRIC PROGRESSION   If $R \neq 1$ then

$$A + AR + AR^2 + \cdots + AR^{n-1} = \frac{A(R^n - 1)}{R - 1} \quad \text{for all positive integers } n.$$

We leave the induction proof to the reader.

DEFINITION   Fibonacci numbers
We define a sequence of numbers as follows:

$$f_1 = 1,$$

$$f_2 = 1,$$

$$f_3 = 1 + 1 = 2,$$

$$f_4 = 1 + 2 = 3,$$

$$f_5 = 2 + 3 = 5,$$

and, in general, if $n \geq 3$, then

$$f_n = f_{n-2} + f_{n-1}.$$

The number f_n is called the nth **Fibonacci number**.

The following table gives the first 16 Fibonacci numbers.

n	1	2	3	4	5	6	7	8	9	10	11	12	13	14	15	16
f_n	1	1	2	3	5	8	13	21	34	55	89	144	233	377	610	987

Various relationships among these numbers may be discovered by trial and error and proved by induction. Let us, for example, add up initial groups of Fibonacci numbers with even subscripts.

$$f_2 = 1$$

$$f_2 + f_4 = 1 + 3 = 4$$

$$f_2 + f_4 + f_6 = 1 + 3 + 8 = 12$$

$$f_2 + f_4 + f_6 + f_8 = 1 + 3 + 8 + 21 = 33$$

$$f_2 + f_4 + f_6 + f_8 + f_{10} = 1 + 3 + 8 + 21 + 55 = 88$$

The reader may notice that 1, 4, 12, 33, and 88 are 1 less than 2, 5, 13, 34, and 89, all Fibonacci numbers with odd subscripts.

THEOREM   For each positive integer n

$$f_2 + f_4 + \cdots + f_{2n} = f_{2n+1} - 1.$$

***Proof** (Induction)*

1. If $n = 1$ this says $f_2 = f_3 - 1$, or $1 = 2 - 1$.
2. Assume $f_2 + f_4 + \cdots + f_{2k} = f_{2k+1} - 1$. Adding $f_{2(k+1)}$ to both sides gives

$$f_2 + f_4 + \cdots + f_{2k+2} = f_{2k+1} + f_{2k+2} - 1 = f_{2k+3} - 1$$

by the definition of the Fibonacci sequence. Since this is $f_{2(k+1)+1}$, we have proved the formula for $n = k + 1$.

Thus by induction the formula is true for all positive integers n.

PROBLEMS FOR SECTION 1.6

A *In problems 1 and 2 identify each sequence as either an Arithmetic Progression (and give A and D) or Geometric Progression (and give A and R), or neither.*

1. a. 2, 4, 8, 16, 32
    b. 9, 7, 5, 3
    c. 3, -3, 3, -3, 3, -3
    d. 1, $\frac{1}{2}$, $\frac{1}{3}$, $\frac{1}{4}$, $\frac{1}{5}$
    e. 1, -2, 3, -4, 5
    f. -5, 1, 7

2. a. 3, 1, $\frac{1}{3}$, $\frac{1}{9}$
    b. $-\frac{1}{2}$, $\frac{1}{2}$, $\frac{3}{2}$, $\frac{5}{2}$
    c. -1, 1, $\frac{1}{2}$, 4
    d. $-\frac{1}{2}$, 1, -2
    e. 1, 4, 9, 16, 25
    f. 0.1, 0.01, 0.001

In problems 3 through 8 compute the sum by using the appropriate formula.

3. $1 + 2 + 3 + \cdots + 100$
4. $5.01 + 5.03 + 5.05 + \cdots + 5.99$
5. $1 + \frac{1}{3} + \frac{1}{9} + \frac{1}{27} + \frac{1}{81} + \frac{1}{243}$
6. $-1 + 2 - 4 + 8 - 16 + \cdots + 512$
7. $\frac{3}{2} + \frac{2}{3} + \frac{8}{27} + \frac{32}{243}$
8. $3 + 2.5 + 2 + \cdots + (-7.5)$

9. Compute f_n for $n = 17$ through 20.
10. Compute $f_1 + f_2 + \cdots + f_n$ for $n = 1, 2, 3, 4, 5, 6$. Conjecture a simpler formula for this sum.

B *In problems 11 through 45 the statement is to be proved true for all positive integers n by induction.*

11. $1 + 4 + 9 + \cdots + n^2 = n(n + 1)(2n + 1)/6$.
12. $1 + 8 + 27 + \cdots + n^3 = n^2(n + 1)^2/4$.
13. $7^n \equiv 6n + 1 \pmod{36}$.
14. $2 \cdot 7^n \equiv 2^n(2 + 5n) \pmod{25}$.
15. $2^n + 3^n \equiv 5^n \pmod 6$.
16. $16^n \equiv 1 - 10n \pmod{25}$.
17. $6 \mid n(n^2 + 3n + 8)$.
18. $6 \mid n^3 - n$.
19. $1 \cdot 1! + 2 \cdot 2! + \cdots + n \cdot n! = (n + 1)! - 1$.
20. $3 \mid 4^n - 1$.
21. $1/(1 \cdot 2) + 1/(2 \cdot 3) + \cdots + 1/n(n + 1) = n/(n + 1)$.
22. $1/(1 \cdot 3) + 1/(3 \cdot 5) + \cdots + 1/(2n - 1)(2n + 1) = n/(2n + 1)$.
23. $2^n \mid (n + 1)(n + 2) \cdots (2n)$.
24. If $a \mid b_i$ for $i = 1, \ldots, n$, and if $x_1, \ldots, x_n$ are integers, then $a \mid x_1 b_1 + x_2 b_2 + \cdots + x_n b_n$.
25. If $a \mid b_1 b_2 \cdots b_n c$, and if $(a, b_i) = 1$ for $i = 1, \ldots, n$, then $a \mid c$.
26. If $a_i \equiv b_i \pmod m$ for $i = 1, \ldots, n$, then $a_1 + \cdots + a_n \equiv b_1 + \cdots + b_n \pmod m$.
27. $f_1 + f_3 + f_5 + \cdots + f_{2n-1} = f_{2n}$.
28. $f_1 + f_2 + \cdots + f_n = f_{n+2} - 1$.
29. $f_n < 2^n$.
30. If A is the matrix $\begin{bmatrix} 1 & 1 \\ 1 & 0 \end{bmatrix}$, then

$$A^n = \begin{bmatrix} f_{n+1} & f_n \\ f_n & f_{n-1} \end{bmatrix} \quad \text{for } n > 1.$$

31. $2^{2^n} + 1 \equiv 5 \pmod{12}$.
32. A set with n elements has exactly 2^n subsets.
33. (For those familiar with calculus) $(d/dx)(x^n) = nx^{n-1}$. Assume only $(d/dx)(x) = 1$ and the product rule for differentiation.
34. $(-4)^n \equiv 1 - 5n \pmod{25}$
35. $(\frac{1}{2} + 1)(\frac{1}{2} + \frac{1}{2})(\frac{1}{2} + \frac{1}{3}) \cdots (\frac{1}{2} + 1/n) = (n + 1)(n + 2)/2^{n+1}$.
36. $2^{2n-1}(n!)^2 \geq (2n)!$.
37. $5^n \equiv 1 + 4n \pmod{16}$.
38. $5^n \equiv 8n^2 - 4n + 1 \pmod{64}$.
39. $8^n \mid (4n)!$.
40. $(2n)! \geq (n!)^2 2^n$.
41. $2^{n+2} \mid (2n + 3)!$.

42. $3^{2n+1} \equiv 2(-1)^{n+1} \pmod 5$.   43. $16^n|(6n)!$.
44. $31|2^{5n} - 1$.   45. $17|2^{8n-4} + 1$.

46. Prove the Geometric Progression formula by induction on n.
47. Prove that $2^n|n(n+1)\cdots(2n)$ for all positive integers n. (This is not the same as problem 23.)

C 48. Let n_0 be a fixed integer, and let $S(n)$ be a statement involving n. Suppose that (1) $S(n_0)$ is true, and (2) whenever $S(k)$ is true for $k \geq n_0$, then $S(k+1)$ is true. Prove that $S(n)$ is true for all $n \geq n_0$.

49. Prove that $(2n)! < (n!)^2 4^{n-1}$ for $n \geq 5$.

50. The **Well-Ordering Principle** states that each nonempty set of positive integers contains a least element. Assume the Well-Ordering Principle and prove from it the Principle of Mathematical Induction. (**Hint:** Suppose (1) and (2) of the induction principle hold. Let n be the least positive integer such that $S(n)$ is false.)

51. Give a formal induction proof of the Euclidean Algorithm.

52. What is wrong with the following proof that all things are the same? We prove by induction on n that any n things are the same. If $n = 1$ this is clear. Assume any k things are the same. Then given $k+1$ things a_1, $a_2, \ldots, a_{k+1}$, we have by the induction hypothesis that $a_1 = a_2 = \cdots = a_k$ and $a_2 = a_3 = \cdots = a_{k+1}$. Thus $a_1 = a_2 = \cdots = a_{k+1}$. Thus by the principle of mathematical induction all things are the same.

53. Prove that if m and n are positive integers, and n is odd, then $(a^n - 1, a^m + 1) = 1$ or 2.

Euclid

LIVED ABOUT 300 B.C.

The Greek mathematican Euclid was the most successful textbook writer of all time, his *Elements* having been used in schools for more than 2000 years. Although this work of 13 volumes is usually thought of as being about geometry, it contains much number theory as well, as we will see. Most of the content of

the *Elements* was created by other mathematicians, but Euclid organized the material in a way that quickly made his book preeminent.

Little is known about the author himself, and in fact he was often confused with another Euclid (Euclid of Megara), portraits of the wrong man even having been published with editions of the *Elements*. (Even the likeness shown here was produced after the mathematician's death.) Our Euclid wrote on many subjects besides mathematics, including music, astronomy, mechanics, and optics. (He thought that rays traveled from a person's eyes to the object of his sight.)

EUCLID

The Euclidean Algorithm was probably known before Euclid, but its first recorded appearance in print is in Book VII of the *Elements*, as follows: "Two unequal numbers being set out, and the less being continually subtracted from the greater, if the number which is left never measures the one before it until a unit is left, the original numbers will be prime to one another." Here instead of subtracting q times a from b all at once, we successively subtract a until a number less than a remains. Then this remainder is subtracted from a, etc. Euclid says that if the last nonzero remainder is 1, then a and b are relatively prime.

Chapter 2

Prime Numbers

2.1 Prime Factorization

DEFINITION   prime, composite

A positive integer is said to be **prime** in case it has exactly two positive divisors. A positive integer having more than two positive divisors is said to be **composite**. The number 1 is neither prime nor composite.

Examples   The primes less than 20 are 2, 3, 5, 7, 11, 13, 17, and 19.

THEOREM   Suppose $n > 1$. Then n can be written as a product of primes.

Proof   If n is prime we are done. (We consider a prime to be a product of one prime.) Otherwise n must have a factor, say d, other than itself and 1. Let $n = dd'$. Clearly $1 < d' < n$ also. We now apply the same argument to d and d' as we did to n. This procedure must end, since the factors grow smaller at each step. But it can stop only when each factor has no positive divisors other than 1 and itself, that is, when each factor is prime.

Illustration of the Proof   Suppose $n = 120$. We know that $120 = 12 \cdot 10$. Now $12 = 3 \cdot 4$, and $4 = 2 \cdot 2$. Thus $12 = 3 \cdot 2 \cdot 2$, and all these factors are prime. Likewise $10 = 2 \cdot 5$, and 2 and 5 are prime. Finally,

$$120 = 3 \cdot 2 \cdot 2 \cdot 2 \cdot 5,$$

and 3, 2, and 5 are prime.

Comment on the Proof The proof given is convincing because we can see how each step will proceed, even though the number of steps and branches involved will depend on the particular value of n. In this respect it is like our first proof of Theorem 1.12 (showing that $a \equiv b$ implies $a^n \equiv b^n$). We suspect that a more formal argument using mathematical induction could be given. This is the case; but a slightly different form of the induction principle must be used.

Mathematical Induction II

Let $S(n)$ be a statement about the positive integer n. Suppose

1. $S(1)$ is true.
2. If for some k $S(1)$, $S(2)$, ..., $S(k)$ are all true, then $S(k + 1)$ is true.

Then $S(n)$ is true for all positive integers n.

Comment on Induction II Of course the difference between Induction II and the form of induction of Section 1.6 is entirely in condition (2). With Induction II we are allowed to assume not only that $S(k)$ is true, but also $S(1)$, $S(2)$, ..., $S(k - 1)$. Since we are allowed to assume more, an Induction II proof is actually easier, at least from a logical standpoint. Induction I (which is what we will call the principle of Section 1.6 when we want to distinguish between the two forms) is, however, formally simpler and suffices for most proofs.

The reader should study the induction proof we now give for the theorem of this section and decide why Induction II is needed in it.

THEOREM Suppose $n > 1$ is an integer. Then n can be written as a product of primes.

Proof *(Induction II)*

1. If $n = 1$ there is nothing to prove, since the theorem only concerns values of $n > 1$.
2. Suppose we know the theorem is true for $n = 1, 2, \ldots, k$. Consider the integer $k + 1$. If $k + 1$ is prime, then we are done. Otherwise

$$k + 1 = d \cdot d',$$

where $1 < d < k + 1$ and $1 < d' < k + 1$. Since d and d' are each at most k, by the induction assumption each can be written as a product of primes. Thus n can also be written as such a product.

Thus by Induction II any $n > 1$ can be written as a product of primes.

As another illustration of an Induction II proof we will prove the following theorem. Recall that f_n is the nth Fibonacci number.

THEOREM   The number of subsets of $\{1, 2, \ldots, n\}$ not containing consecutive integers is f_{n+2}.

Example   If $n = 5$ the subsets we want to count are the empty set, $\{1\}$, $\{2\}$, $\{3\}$, $\{4\}$, $\{5\}$, $\{1, 3\}$, $\{1, 4\}$, $\{1, 5\}$, $\{2, 4\}$, $\{2, 5\}$, $\{3, 5\}$, and $\{1, 3, 5\}$, a total of 13 subsets. But $f_{5+2} = f_7 = 13$.

Proof of the theorem (Induction II)

1. If $n = 1$ then the set in question is $\{1\}$, and the subsets we want to count are the empty set and $\{1\}$ itself. But $f_{1+2} = f_3 = 2$, so the theorem holds.
   We will also check the case $n = 2$ so that we may assume that $k > 1$ in the second part of this proof. The subsets of $\{1, 2\}$ not containing consecutive integers are the empty set, $\{1\}$, and $\{2\}$. But $f_{3+2} = f_5 = 3$.
2. Suppose we know that for some integer $k > 1$ the theorem holds for each set $\{1, 2, \ldots, j\}$ for $j = 1, 2, \ldots, k$. Consider the set $\{1, 2, \ldots, k, k+1\}$. Suppose that S is a subset of this set not containing consecutive integers.

Case 1: $k + 1$ is not in S.
   Then S is just a subset of $\{1, 2, \ldots, k\}$ not containing consecutive integers. By the induction hypothesis (for $n = k$) there are exactly f_{k+2} such subsets S.

Case 2: $k + 1$ is in S.
   Then k cannot be in S. In fact if we remove $k + 1$ from S we have left a subset of $\{1, 2, \ldots, k - 1\}$ not containing consecutive integers. Moreover, adding $k + 1$ to any such subset produces a set of the type we are trying to count in this case. By the induction hypothesis for $n = k - 1$ there are exactly $f_{k-1+2} = f_{k+1}$ subsets of $\{1, 2, \ldots, k - 1\}$ not containing consecutive integers.
   Putting the two cases together we find that there are $f_{k+2} + f_{k+1}$ subsets S to count. But by the definition of the Fibonacci numbers this is f_{n+3}, which is what we wanted to prove.

   Thus by Induction II the theorem holds for all positive integers n.

The Number of Divisors of an Integer

It will turn out that the factorization of an integer into primes will tell us how to count its divisors.

DEFINITION $d(n)$
   We define $d(n)$ to be the number of positive divisors of n.

Examples

n	positive divisors of n	$d(n)$
1	1	1
2	1, 2	2
3	1, 3	2
4	1, 2, 4	3
5	1, 5	2
6	1, 2, 3, 6	4
7	1, 7	2
8	1, 2, 4, 8	4
9	1, 3, 9	3
10	1, 2, 5, 10	4

We will examine the function $d(n)$ in more detail in the next section.

PROBLEMS FOR SECTION 2.1

1. What are the primes between 20 and 40?
2. What are the primes between 40 and 60?
3. What are the primes between 100 and 120?
4. What are the primes between 120 and 140?
5. Compute $d(n)$ for n between 11 and 20.
6. Compute $d(n)$ for n between 21 and 30.

In the next four problems write n as a product of primes. Write the primes in nondecreasing order. For example, $120 = 2 \cdot 2 \cdot 2 \cdot 3 \cdot 5$.

   7. $n = 180$ 8. $n = 500$ 9. $n = 1001$ 10. $n = 4199$

In the next four problems compute d(n) for each given n.

   11. $n = 75$ 12. $n = 99$ 13. $n = 243$ 14. $n = 1024$

15. Compute f_3 using the formula of problem 32.
16. Compute f_4 using the formula of problem 32.
17. What do you think $d(pq^2)$ is, given that p and q are distinct primes?
18. What do you think $d(p^2q^3)$ is, given that p and q are distinct primes?
19. What is $d(2^n)$?
20. What is $d(2^n 3^m)$?
21. Find the smallest $n > 0$ such that $d(n) = 5$.
22. Find the smallest $n > 0$ such that $d(n) = 6$.
23. Find an n such that $d(n) = 100$.
24. Find an n not a power of a prime such that $d(n) = 51$.

In the next eight problems prove the statement given is true for all positive integers n.

25. $f_1^2 + f_2^2 + \cdots + f_n^2 = f_n f_{n+1}$
26. $f_n + f_{n+1} + f_{n+3} = f_{n+4}$
27. If t_m is the number of subsets of $\{1, 2, \ldots, n\}$ not containing three consecutive integers, then $t_{n+4} = 2t_{n+3} - t_n$.
28. $f_1 - f_2 + f_3 - \cdots + (-1)^n f_{n+1} = (-1)^n f_n + 1$
29. If $0 < k < n$, then $f_{n+1} = f_{n-k} f_k + f_{n-k+1} f_{k+1}$. (**Hint:** Fix k and use Induction II on n starting with $n = k + 1$.)
30. $f_n < (5/3)^n$
31. $2^n f_n < (\sqrt{5} + 1)^n$
32. $f_n = (A^n - B^n)/\sqrt{5}$, where A and B are the positive and negative solutions to $x^2 - x - 1 = 0$, respectively.

33. Let $a_1 = 1$, $a_2 = 3$, and $a_n = a_{n-1} + a_{n-2}$ for $n = 3, 4, \ldots$. Show that $a_n < (7/4)^n$ for all positive integers n.
34. Prove that if p is prime, then $d(p^n) > n$.
35. Suppose that p and q are distinct primes, and that p and q are the only primes dividing pq. Show that $d(pq) = 4$.
36. Show that Induction II implies Induction I.
37. Show that Induction I implies Induction II. (**Hint:** Suppose $S(n)$ is a statement satisfying (1) and (2) of Induction II. Let $S'(n)$ be the statement "$S(k)$ is true for $k = 1, 2, \ldots, n$." Apply Induction I to $S'(n)$.)
38. Show that Induction II implies the Well-Ordering Principle. (See the problems at the end of Section 1.6.) (**Hint:** Let T be a set of positive integers with no smallest element. Let $S(n)$ be the statement that n is not in T. Show that T is empty.)
39. Show that the sum of two consecutive primes is never twice a prime.

2.2 Factorization in Other Systems

What Is d(pq)?

Recall that $d(n)$ denotes the number of positive divisors of n. It is easy enough for us to compute that $d(2 \cdot 3) = 4$, since the positive divisors of 6 are 1, 2, 3, and 6. In general, if p and q are distinct primes then pq has the divisors 1, p, q, and pq; and so $d(pq)$ is at least 4. The reader may feel it obvious that $d(pq)$ is exactly 4. After all, what other divisors could pq have?

This is a situation where our familiarity with the integers may be a disadvantage, since it may lead us to assume as true things that really should be proved. It happens to be true that pq has no factors other than those listed, but we cannot let intuition and experience substitute for a rigorous proof of this fact. As useful as intuition is in mathematics, there are many instances where it has led to serious errors.

In order to demonstrate that one should not put too much trust into one's intuition about the integers, we will now present two examples of systems in which certain laws with which we are familiar do not hold.

The System E

Let E be the set of all even positive integers: $2, 4, 6, 8, \ldots$. We say that a **divides** b in E in case $b = ak$ for some element k in E. Of course this parallels the usual definition of divisibility in the integers. For example 2 divides 8 in E since $8 = 2 \cdot 4$, and 4 is in E. On the other hand 2 does not divide 6 in E because although $6 = 2 \cdot 3$, 3 is not in E.

Since 1 is not in E we will have to revise our definition of primality; but the definition we will give will be equivalent to the previous one in the case of the ordinary integers. We will say that p in E is **prime** if p cannot be written as the product of two other elements of E greater than 1; for example 2 and 6 are prime in E, but $4 = 2 \cdot 2$ is not.

The reader may easily check that 10 and 30 are also prime in E. Thus $60 = 6 \cdot 10$ is of the form pq, where p and q are distinct primes.

Now comes the shocker.

The integer $60 = 2 \cdot 30$, and 2 and 30 are primes. We have factored 60 into prime factors in two essentially different ways, something (it will turn out) we could never do with the ordinary integers.

Note that the system E is not at all exotic, and in fact shares most of the usual algebraic properties of the integers. For example it is closed under both addition and multiplication. One basic property it lacks is that of having a multiplicative unit 1. Our next example will contain 1.

The System T

Let T be all positive integers congruent to 1 modulo 3; that is, the numbers $1, 4, 7, 10, \ldots$. Since $a \equiv 1 \pmod 3$ and $b \equiv 1 \pmod 3$ imply $ab \equiv 1 \pmod 3$ we see T is closed under multiplication.

In T we say a **divides** b in case $b = ak$ for some k in T, and call an element $p > 1$ of T **prime** if it cannot be written as the product of two factors in T greater than 1.

We find the smallest composite element of T to be $16 = 4 \cdot 4$, and it is easy to check that 4, 10, 22, and 55 are all prime elements of T.

Another shocker.

For these numbers provide an example of nonunique prime factorization in T, since $220 = 4 \cdot 55 = 10 \cdot 22$. In fact 220 has the six divisors 1, 4, 10, 22, 55, and 220 in T, in spite of the fact that 220 is of the form pq for distinct primes p and q.

Chastened by these examples, we will prove a theorem that will pave the way toward showing that such aberrations cannot occur among the ordinary integers.

THEOREM 2.1   Suppose p is prime and $p \,|\, ab$. Then $p \,|\, a$ or $p \,|\, b$.

Proof   Suppose $p \nmid a$. Then, since the only positive divisors of p are p and 1, (a, p) must be 1. Thus by Theorem 1.5 $p \,|\, b$.

PROBLEMS FOR SECTION 2.2

1. List the primes in E between 1 and 20.
2. List the primes in E between 21 and 40.
3. List the primes in T between 1 and 30.
4. List the primes in T between 31 and 60.

In the next four problems find the number of divisors of n in E.

5. $n = 40$ 6. $n = 60$ 7. $n = 90$ 8. $n = 84$

In the next four problems find the number of divisors of n in T.

9. $n = 64$ 10. $n = 85$ 11. $n = 250$ 12. $n = 550$

13. How many divisors does a prime in E have?
14. How many divisors does a prime in T have?
15. Find a number in T less than 220 that can be factored into primes in two essentially different ways.
16. Show that T is not closed under addition.

17. The set T consists of all positive integers of the form $3k + 1$. Use this to prove without using congruences that T is closed under multiplication.
18. Prove that E is closed under addition and multiplication.
19. Use Theorem 2.1 to show that if p and q are distinct primes, then $d(pq) = 4$.
20. Show by induction that if a prime p divides a product of n numbers, then it divides at least one of the numbers.
21. Prove that if n is in E then n can be written as a product of primes in E. Use Induction II.

2.3   The Fundamental Theorem of Arithmetic

Fermat's Last Theorem

The assumption that the factorization of a number into primes must be unique has a famous precedent in mathematical history. Consider the equation

$$x^n + y^n = z^n.$$

This is an example of a **Diophantine equation**, that is, an equation for which integral (or, sometimes, rational) solutions are desired. The name comes from the Greek mathematician Diophantus, who studied many such equations.

If $n = 2$ it is easy to find solutions to our equation; for example $x = 3$, $y = 4$, $z = 5$, or $x = 12$, $y = 5$, $z = 13$. For $n > 2$, however, it is a different story; and not one solution in positive integers is known.

Around 1637 the French mathematician Pierre de Fermat, writing in the margin of his copy of the works of Diophantus, claimed that he had found a "truly wonderful" proof that the equation has no positive solutions for $n > 2$, but that there was not room for him to write it down there. This statement is known as **Fermat's Last Theorem**. If Fermat wrote down his alleged proof anywhere, no one ever found it, nor were later mathematicians able to construct a proof.

More than 200 years after Fermat's claim interest developed in the connection between Fermat's Last Theorem and algebraic integers. These are real or complex numbers that behave like the ordinary integers in many respects. In particular, the concepts of divisibility and primality can be defined in various sets of algebraic integers. The French mathematicians Lamé and Cauchy thought they were close to proofs of Fermat's Last Theorem, based on their assumption that prime factorization of algebraic integers was unique. An often-repeated story is that the German mathematician Kummer submitted a manuscript purporting to contain a proof, which was invalid because of this assumption. (Inside many classes of algebraic integers factorization is *not* unique; the set S that we will define shortly is one example.) Kummer's supposed manuscript has never been found, however, and doubt has been thrown on this part of the story. The interested reader should consult the book by Edwards listed in the references.

In any case Kummer went on to prove many important results related to Fermat's Last Theorem, including the impossibility of the equation for a large number of values of n. His work led the way to the modern theory of algebraic numbers.

The System S

We define S to be the set of all complex numbers of the form

$$a + b\sqrt{-6},$$

where a and b are ordinary integers. It is easy to see that S is closed under addition and, since

$$(a + b\sqrt{-6})(c + d\sqrt{-6}) = ac - 6bd + (ad + bc)\sqrt{-6},$$

also under multiplication.

As usual, we say $A|B$ in S in case $B = AC$ for some element C of S. Since the concepts of being positive or negative do not apply to complex numbers, we must modify our definition of primality. We say P in S is **prime** in case we cannot write P as a product AB where A and B are in S and neither A nor B is 1 or -1.

It is perhaps not clear that elements of S can be factored into primes, since the argument given in Section 2.1 depended on the factors of a number being smaller than the number; while here the factors are complex numbers. To get around this, we define a function N on S by

$$N(a + b\sqrt{-6}) = a^2 + 6b^2.$$

We call $N(A)$ the **norm** of A. It is not hard to check that

1. $N(A) = 0$ if and only if $A = 0$;
2. $N(A) = 1$ if and only if $A = 1$ or -1;
3. $N(AB) = N(A)N(B)$.

(These and many other details concerning S will be left for the exercises at the end of this section.)

Then $N(A)$ is a nonnegative integer for A in S, and if C is not prime in S then $C = AB$, where

$$1 < N(A) < N(C) \qquad \text{and} \qquad 1 < N(B) < N(C).$$

From this it is easy to show that any element of S other than 0, 1, and -1 has a prime factorization.

As a sample of the use of the function N we will show that 2 is prime in S. Suppose $2 = AB$, where neither A nor B is 1 or -1. Then

$$N(A)N(B) = N(AB) = N(2) = 4,$$

by property (3). Since $N(A)$ and $N(B)$ are both ordinary integers > 1 we must have $N(A) = 2$. Let $A = a + b\sqrt{-6}$. Then $a^2 + 6b^2 = 2$. This is impossible whether $b = 0$ or b is not 0.

It may be shown that S does not enjoy the unique factorization property. The proof is left for the exercises.

The Unique Factorization Property

Now we return to the ordinary integers and show that they do enjoy the unique factorization property. Suppose $n > 1$ has two factorizations into primes, say

$$n = p_1 p_2 \cdots p_s = q_1 q_2 \cdots q_t.$$

Since p_1 divides the right side of the last equation, Theorem 2.1 implies that p_1 divides either q_1 or $q_2 \cdots q_t$. In the latter case p_1 divides either q_2 or the product of the remaining q's by the same argument. Eventually we find that p_1 divides one of the q's, say q_i. But since q_i is prime, its only divisors are itself and 1, so we must have $p_1 = q_i$.

Now we divide both sides of the equation by $p_1 = q_i$ and apply the same argument to p_2, finding that it must equal one of the remaining q's. We cancel this prime in the same way. This process is continued as long as possible. Both sides must be exhausted at the same time, otherwise we would have a product of primes equaling 1. We have matched each p with a q. Thus the q's must be a simple rearrangement of the p's.

Notice that the key to proving unique factorization is Theorem 2.1. This result may be found in Book VII of Euclid's *Elements*, stated as follows: "If two numbers by multiplying one another make some number, and any prime number measure the product, it will also measure one of the original numbers."

The theorem just proved is often combined with the theorem giving the existence of a prime factorization, as follows:

THE FUNDAMENTAL THEOREM OF ARITHMETIC   Any integer $n > 1$ has a factorization into primes. This factorization is unique up to the order of the factors.

DEFINITION   divides exactly
Suppose p is a prime and $k > 0$. We say p^k **divides** a **exactly**, and write $p^k \| a$, in case $p^k | a$ but $p^{k+1} \nmid a$.

Examples $3 \| 6$, $2^5 \| 96$, $9 \| -18$, $9 \| 27$ is false, $4 \| 24$ is false, $4 \| 62$ is false.

PROBLEMS FOR SECTION 2.3

Problems 1 through 10 refer to the set S defined in this section.

1. Find $N(7 + 3\sqrt{-6})$ 2. Find $N(3 - 7\sqrt{-6})$

In the next four problems find B and write it in the form $a + b\sqrt{-6}$.

3. $(3 - 7\sqrt{-6})B = 147 - 40\sqrt{-6}$
4. $-200 + 19\sqrt{-6} = (2 + 3\sqrt{-6})B$
5. $(-7 + 2\sqrt{-6})B = 73$
6. $(3 + \sqrt{-6})B = -15$

In the next four problems the number A given is not prime. List all factors $B = a + b\sqrt{-6}$ with $1 < N(B) < N(A)$ and a and b positive.

7. $A = 127$ 8. $A = 193$ 9. $A = 79$ 10. $A = 103$

B | *Problems 11 through 19 refer to the set S defined in this section.*

11. Show that $N(A) = 0$ if and only if $A = 0$, and $N(A) = 1$ if and only if A is 1 or -1.
12. Suppose P is prime in S. How many divisors does P have?
13. Is 7 prime in S? List all its divisors.
14. Is 31 prime in S? List all its divisors.
15. Show S is closed under addition.
16. Show that if A divides 1, then A is 1 or -1.
17. Show that 5 is prime in S.
18. Show that $2 + \sqrt{-6}$ and $2 - \sqrt{-6}$ are prime in S.
19. Show that the prime factorization of 10 in S is not unique.

True-False. In the next 11 problems tell which statements are true and give counterexamples for the others. Assume p is prime, a and b are nonzero integers, i, j, and k are positive integers, and $i < j$.

20. If $p^i \| a$ and $p^j \| b$, then $p^{i+j} \| ab$.   21. If $p^i \| (a, b)$, then $p^{2i} \| ab$.
22. If $p^i \| a$ and $(a, b) = 1$, then $p^i \| ab$.
23. If $p^i \| a$ and $(a, b) > 1$, then $p^i \| ab$ is false.
24. If $p^j \| a$ and $p^i \| (a, b)$, then $p^{j+i} \| ab$.
25. If $p^i \| a$ and $p^j \| (a, b)$, then $a + b$ is prime.
26. If $p^i \| a$, then $p^{ki} \| a^k$.   27. $(a^k, b^k) = (a, b)^k$.
28. If $p^i \| a$ and $p^i \| b$, then $p^i \| a + b$.   29. If $p^i \| a$ and $p^j \| b$, then $p^i \| a + b$.
30. If $p^i \| a$ and $p^i \| b$, then there exists $m \geq i$ such that $p^m \| a + b$.

31. Give a solution to $x^2 + y^2 = z^2$ in positive integers other than the two given in this section.
32. When 2^{33} is computed on a certain hand calculator the answer 8589934605 is given. What does this have to do with the Fundamental Theorem of Arithmetic?

C | *Problems 33 through 36 refer to the set S.*

33. Show that $N(AB) = N(A)N(B)$ in S using the definition of N.
34. Show that $N(A) = |A|^2$, where $|A|$ is the modulus of the complex number A. Use this to show $N(AB) = N(A)N(B)$.
35. Prove that if A in S is not 0, 1 or -1, then A has a prime factorization. Use Induction II on $n = N(A)$.
36. Show that if $N(A)$ is a prime in the ordinary sense, then A is prime in S.

37. Show that $a|b$ if and only if whenever p is prime and $p^k|a$, then $p^k|b$.
38. Use the previous problem to show that if a and b are positive integers, then $(a, b) = (a + b, [a, b])$.
39. Why is the concept of dividing exactly only applied to powers of primes?
40. Show that if $x^2 + y^2 = z^2$ and if no prime divides all three of x, y, and z, then z is odd.

41. Show that if it could be proved that $x^n + y^n = z^n$ was impossible in positive integers for $n = 4$ and for n any odd prime, then Fermat's Last Theorem would be proved.

42. Very little is known of the life of Diophantus. A collection of problems published in the fifth or sixth century contains the following problem about him:

"This tomb holds Diophantus. Ah, what a marvel! And the tomb tells scientifically the measure of his life. God vouchsafed that he should be a boy for the sixth part of his life; when a twelfth was added, his cheeks acquired a beard; He kindled for him the light of marriage after a seventh, and in the fifth year after his marriage He granted him a son. Alas! late-begotten and miserable child, when he had reached the measure of half his father's life, the chill grave took him. After consoling his grief by this science of numbers for four years, he reached the end of his life."

How many years did Diophantus live?

43. Prove that if $n > 1$, then $n = mk^2$, where the square of no prime divides m.

44. Show that if $a^2 + b^2 = c^2$, then $3 \mid abc$.

45. Prove that if $n = 2^k$, $k > 0$, then $2^{k+2} \parallel 3^n - 1$.

46. Prove that if $n = 2^k t$, $k > 0$, t odd, then $2^{k+2} \parallel 3^n - 1$.

2.4 Prime Power Factorizations

The Importance of the Fundamental Theorem

The existence and uniqueness of the prime decomposition of an integer are very important for understanding the integers and proving things about them. Thinking about an integer in terms of its prime factorization provides an entry to many number-theoretic problems. When a proof using prime factorization works, it is usually straightforward, if not necessarily elegant. The next theorem shows how the concepts of divisibility and of the gcd and lcm depend on prime factorization.

THEOREM 2.2 Suppose a and b are positive integers. Let the distinct primes dividing a or b (or both) be $p_1, p_2, \ldots, p_n$. Suppose

$$a = p_1^{j_1} p_2^{j_2} \cdots p_n^{j_n} \qquad \text{and} \qquad b = p_1^{k_1} p_2^{k_2} \cdots p_n^{k_n}.$$

(Some of the j's and k's may be 0.) Let m_i be the smaller and M_i be the larger of j_i and k_i for $i = 1, 2, \ldots, n$. Then

a. $a \mid b$ if and only if $j_i \le k_i$ for $i = 1, 2, \ldots, n$.

b. $(a, b) = p_1^{m_1} p_2^{m_2} \cdots p_n^{m_n}$.

c. $[a, b] = p_1^{M_1} p_2^{M_2} \cdots p_n^{M_n}$.

Proof *(Partial)*

a. Suppose $a|b$. Then $b = ac$ for some integer c. Consider the prime factorizations of a, b, and c. According to the equation $b = ac$ every prime appearing in the factorization of a must also appear in that of b, and at least as many times. (Here we have used the fact that the factorization is unique up to order.) Thus $j_i \le k_i$ for all i.

Since part (a) contains an "if and only if" statement, this is only half of its proof. We leave the other half for the exercises at the end of this section.

b. Since $m_i \le$ both j_i and k_i for all i, part (a) shows the expression on the right side of (b) to be a common divisor of a and b.

Now suppose d is any common divisor of a and b. Then by part (a) again $d = p_1^{r_1} p_2^{r_2} \cdots p_n^{r_n}$, where $r_i \le j_i$ and $r_i \le k_i$ for all i. Then $r_i \le m_i$ for all i and so d is less than or equal to the right side of (b). Thus the right side of (b) is a common divisor of a and b exceeding any other common divisor. By definition it is (a, b).

c. This proof is left for the problems.

Examples of the Use of Theorem 2.2 Suppose $a = 75 = 3^1 5^2$, $b = 900 = 2^2 3^2 5^2$, and $c = 440 = 2^3 5^1 11^1$. Then

a. $a|b$, since $a = 2^0 3^1 5^2$, and $0 \le 2$, $1 \le 2$, and $2 \le 2$;
b. $(b, c) = 2^2 3^0 5^1 11^0 = 20$;
c. $[b, c] = 2^3 3^2 5^2 11^1 = 19800$.

Factoring Large Numbers

Using part (b) of the last theorem appears to be easier than using the Euclidean Algorithm. This method depends on knowing the prime factorizations of the numbers involved, however. If n has mostly small prime factors, the factorization of n may be found by dividing through by each factor as it is found. If n has only large prime divisors, however, finding them may be difficult. In such a case the Euclidean Algorithm and the equation $[a, b] = ab/(a, b)$ will still provide the easiest way to find the gcd and lcm of two numbers.

Very often the digits of a number tell us something about its divisors. For example, we know the number 7586634 is even without bothering to divide it by 2—because its last digit is even. This idea is so familiar that we have probably never thought about why it is true. Writing out a formal proof may be instructive.

THEOREM   A positive integer written to base 10 is even if and only if its last digit is even.

Proof   Suppose n has the digits $a_k, a_{k-1}, \ldots, a_0$, starting from the left, so

$$n = a_k 10^k + \cdots + a_1 10^1 + a_0.$$

We want to show that n and a_0 are either both even or both odd. Another way to say this is that

$$n \equiv a_0 \pmod 2.$$

But this is implied by the fact that $10 \equiv 0 \pmod 2$ and what we know about manipulating congruences (Theorems 1.8 and 1.11).

Properties of Digits

The key to the proof just given is the fact that $10 \equiv 0 \pmod 2$, and we write numbers in terms of powers of 10 (no doubt because humans have ten fingers). If a different base were used the theorem might not be true.

Similar theorems can be proved for other divisors d, provided that 10 is sufficiently simple modulo d. For example, the same proof shows that a number is divisible by 5 if and only if its last digit is. Actually, the proof says even more, namely that a positive integer is congruent modulo 5 to its last digit. Thus we can say not only that 38707 is not divisible by 5, but also that its least residue modulo 5 is 2, since this is true for 7.

A similar proof shows that a positive integer is congruent to the sum of its digits modulo both 3 and 9. In particular, n is divisible by 3 (or 9) if and only if the sum of its digits is. We will do the proof only for 3.

THEOREM   Any positive integer is congruent to the sum of its digits modulo 3.

Proof   Let $n = a_k 10^k + \cdots + a_0$. Note that $10 \equiv 1 \pmod 3$. Then

$$n \equiv a_k (1)^k + \cdots + a_1(1) + a_0 \pmod 3.$$

A similar theorem holds for the modulus 11. We omit the proof.

THEOREM   If $n = a_k 10^k + \cdots + a_0$, then

$$n \equiv a_0 - a_1 + a_2 - \cdots \pmod{11}.$$

The following summarizes our results on digits.

THE DIGIT THEOREM   Let $n > 0$ have the decimal representation $a_k a_{k-1} \cdots a_0$. Then

a. $n \equiv a_0 \pmod 2$
b. $n \equiv a_0 \pmod 5$
c. $n \equiv a_k + \cdots + a_0 \pmod 3$
d. $n \equiv a_k + \cdots + a_0 \pmod 9$
e. $n \equiv a_0 - a_1 + \cdots \pmod{11}$.

Example We will factor $n = 37719$. Since $3 + 7 + 7 + 1 + 9 = 27$, we see $9 \mid n$. Division produces $n = 9 \cdot 4191$. Now we concentrate on 4191. Since $4 + 1 + 9 + 1 = 15$, 4191 is divisible by 3 but not 9. We have $4191 = 3 \cdot 1397$. Ordinary division shows $7 \nmid 1397$. The next prime is 11. This is a factor since $7 - 9 + 3 - 1 = 0$ and $11 \mid 0$. In fact $1397 = 11 \cdot 127$. Since 11 doesn't divide $7 - 2 + 1$, we go on to 13.

Now $13^2 = 169$, so $127/13 < 13$. Thus if 127 had a proper factor ≥ 13, then it would also have a factor < 13. Since the latter possibility has been eliminated, we conclude that 127 is a prime. Thus $37719 = 3^3 \cdot 11 \cdot 127$.

THEOREM 2.3 If the integer $n > 1$ has no prime divisor $\leq \sqrt{n}$, then n is prime.

Proof This proof is illustrated by the argument about 13 at the end of the last example. Suppose n is composite. Then $n = d_1 d_2$, where both d_1 and d_2 exceed 1. If both d_1 and d_2 exceed $\sqrt{n}$, then

$$n = d_1 d_2 > (\sqrt{n})^2 = n,$$

which is impossible. Suppose $d_1 \leq \sqrt{n}$. Then d_1 is either prime or else has a prime divisor $\leq \sqrt{n}$.

PROBLEMS FOR SECTION 2.4

 *In the first ten problems write each integer in the form $p_1^{k_1} p_2^{k_2} \cdots p_n^{k_n}$, where the p's are distinct primes, the k's are positive integers, and $p_1 < p_2 < \cdots < p_n$. This is called the **prime power factorization** of the integer.*

1. 293	2. 1001	3. 1763	4. 2310
5. 6561	6. 1000000	7. 1111	8. 16281
9. 14641	10. 44032		

Find the largest power of 3 dividing n in each of the next four problems.

11. $n = 2139$ 　　　12. $n = 142155$ 　　　13. $n = 57035$ 　　　14. $n = 11799$

In each of the next six problems use the prime factorization of a and b (a) to decide whether $a \mid b$, (b) to compute (a, b), and (c) to compute $[a, b]$.

15. $a = 495, b = 3861$ 　　　　　　16. $a = 4600, b = 2116$
17. $a = 266, b = 748$ 　　　　　　18. $a = 1287, b = 9009$
19. $a = 2090, b = 1911$ 　　　　　　20. $a = 2592, b = -3888$

B *In the next four problems, suppose that (to base 10) n consists of the digit d repeated k times. If $d = 3$ and $k = 5$, for example, then $n = 33333$.*

21. When does 11 divide n? 　　　　22. When does 3 divide n?
23. When does 9 divide n? 　　　　24. When does 2 divide n?

25. Prove that any positive integer is congruent to its last digit modulo 5.
26. Prove that any positive integer is congruent to the sum of its digits modulo 9.
27. What is the least residue of 94850372 modulo 9? Modulo 11?
28. What is the least residue of 1280054 modulo 9? Modulo 11?
29. What is the least residue of -85763 modulo 3? Modulo 11?
30. What is the least residue of -158473 modulo 5? Modulo 11?
31. What is the least residue of 10^{99} modulo 9? Modulo 11?
32. What is the least residue of 17615^3 (mod 11)?
33. What is the least residue of 65432^3 (mod 9)?
34. What is the least positive integer k that makes the following statement true: "If $1 < n < 1000$ and if n has no positive divisor $< k$, then n is prime"?
35. What is the least positive integer k that makes the following statement true: "If $1 < n < 2000$ and if n has no positive divisor $< k$, then n is prime"?
36. Prove part (e) of the Digit Theorem.
37. Show that in the notation of the Digit Theorem $n \equiv 10a_1 + a_0$ (mod 4).
38. A calculator says $2^{31} = 2147483654$. Show this is wrong.

C

39. Give a test for the divisibility of n by 3 depending on the digits of n when written to base 9.
40. Give a test for the divisibility of n by 2 depending on the digits of n when written to base 9.
41. Show that if n is a positive integer, then n is divisible by 2^k if and only if the number formed by dropping all but the last k digits of n is also.
42. Give the rest of the proof for part (a) of Theorem 2.2.
43. Prove part (c) of Theorem 2.2.
44. Give a proof of Theorem 1.5 using the Fundamental Theorem of Arithmetic.
45. Show that a and b are relatively prime if and only if whenever p is prime and $p|a$, then $p \nmid b$.
46. Give a proof of the last theorem of Section 1.2 using Theorem 2.2.
47. Prove that if n^2 and $2n^2$ are written in base 3, then their last nonzero digits are 1 and 2, respectively ($n > 0$).
48. Use the last problem to prove that $\sqrt{2}$ is irrational.
49. Show that $111 = k^2$ cannot be correct in any base.
50. Suppose $n > 0$ has the property that whenever $p|n$, p prime, then $p^2|n$. Show that there exist integers a and b such that $n = a^3b^2$.

2.5 The Set of Primes Is Infinite

The Sieve of Eratosthenes

Suppose we want to determine all the primes less than 100. We might proceed as follows. First we write out the integers from 2 to 100. We know 2 is prime; let us circle it and cross out the remaining even numbers on our list. The next lowest number that hasn't been crossed out is 3; we circle it and cross out every third number thereafter, since 3 is a proper divisor of each of these. Our list now looks as follows:

② ③ ̶4̶ 5 ̶6̶ 7 ̶8̶ ̶9̶ ̶1̶0̶ 11 ̶1̶2̶ 13 ̶1̶4̶ ̶1̶5̶ ̶1̶6̶ 17 ···

At each stage in this procedure the smallest number that has not been circled or crossed out must be prime, since otherwise it would have a smaller prime divisor, and so have been eliminated already.

This process is called the **Sieve of Eratosthenes**, after the Greek scientist who invented it. Note that according to Theorem 2.3, in order to find all the primes up to n we need only sieve out multiples of primes $\leq \sqrt{n}$. To find the primes up to 100, for example, we need only cross out multiples of 2, 3, 5, and 7.

The operation of the sieve of Eratosthenes suggests that primes should be rarer among the larger integers. For example, its application to the numbers between 100 and 150 consists in crossing out the multiples of the 5 primes $\leq \sqrt{150} = 12.2 \ldots$. Applying the sieve between 1000 and 1050, however, we eliminate not only the multiples of all these primes, but also the multiples of the additional 6 primes between 12.2 and $\sqrt{1050} = 32.4 \ldots$. It turns out that there are 10 primes between 100 and 150, but only 8 between 1000 and 1050. Between 10000 and 10050 there are just 4 primes.

These considerations suggest the possibility that at some stage in the application of the sieve all larger numbers will have been crossed out, so there would be no more primes. This would mean that there would exist only finitely many primes, say $p_1, p_2, \ldots, p_k$, so that each integer greater than 1 could be written as a product of powers of these primes. The integer

$$P = p_1 p_2 \cdots p_k$$

would then have the interesting property that $(P, n) > 1$ whenever $n > 1$, since all the distinct prime factors of n would appear in P. Let us make a table of (P, n) for small values of n.

n	1	2	3	4	5	6	7	8	9	10
(P, n)	1	2	3	2	5	6	7	2	3	10

Back in Section 1.5 we made a similar table for $(6, n)$. It turned out to be periodic, with period 6. In fact we proved in Theorem 1.10 that if $b \equiv b' \pmod{a}$, then $(a, b) = (a, b')$. But this doesn't square with what we know about (P, n). Theorem 1.10 says that if $n \equiv 1 \pmod{P}$, then $(P, n) = (P, 1)$; for example $(P, P + 1) = 1$. Our definition of P, however, led us to the conclusion that $(P, n) > 1$ for $n > 1$. The trouble must be in our assumption that the number of primes is finite, which has produced a contradiction. Thus we have proved the following theorem:

THEOREM 2.4 The number of primes is infinite.

Euclid's Proof That the Number of Primes Is Infinite   Theorem 2.4 first appears in the works of Euclid, so for historical interest we will give Euclid's proof, which has the advantage of depending on little beyond the definition of the primes.

As before, we assume there are only the primes $p_1, p_2, \ldots, p_k$ and no more. Consider the number

$$Q = p_1 p_2 \cdots p_k + 1.$$

Now Q is either prime or else has a prime factor. If Q is prime, we have a contradiction, since $p_1, p_2, \ldots, p_k$ are supposed to be all the primes. But any prime factor of Q must be different from all of $p_1, p_2, \ldots, p_k$, since it is easy to see that none of the p's can divide Q. This again contradicts the assumption that $p_1, p_2, \ldots, p_k$ comprise all the primes.

Euler's Proof That the Number of Primes Is Infinite   Now we give a more sophisticated proof of Theorem 2.4 by the Swiss mathematician Leonhard Euler. Some exposure to infinite series is desirable (but not absolutely necessary) for its understanding. Again we assume $p_1, p_2, \ldots, p_k$ are all the primes. Then if n is any positive integer we may choose r big enough so that the term $1/n$ appears when we multiply out the product

$$\left(1 + \frac{1}{p_1} + \frac{1}{p_1^2} + \cdots + \frac{1}{p_1^r}\right)\left(1 + \frac{1}{p_2} + \frac{1}{p_2^2} + \cdots + \frac{1}{p_2^r}\right) \cdots \left(1 + \frac{1}{p_k} + \frac{1}{p_k^2} + \cdots + \frac{1}{p_k^r}\right).$$

In fact increasing r merely adds in more terms. For example we get $1/12$ by choosing $1/2^2$ from the first factor, $1/3$ from the second, and 1 from all the others (assuming that $p_1 = 2$ and $p_2 = 3$).

Now by the formula for the sum of a geometric progression (see Section 2.6)

$$1 + \frac{1}{p} + \frac{1}{p^2} + \cdots + \frac{1}{p^r} = \frac{1 - p^{-r-1}}{1 - p^{-1}} < \frac{1}{1 - p^{-1}} = \frac{p}{p-1}.$$

We see that for any n the sum

$$1 + \frac{1}{2} + \frac{1}{3} + \cdots + \frac{1}{n}$$

cannot exceed

$$\frac{p_1}{p_1 - 1} \cdot \frac{p_2}{p_2 - 1} \cdot \ldots \cdot \frac{p_k}{p_k - 1}.$$

But this contradicts the divergence of the harmonic series $1 + 1/2 + 1/3 + \cdots$. (To get the contradiction without knowing about the harmonic series, see Problem 14 at the end of this section.)

Other Theorems Concerning Primes

The distribution of primes has been studied extensively and is a central topic in what is known as **analytic number theory**. Unfortunately what can be proved mostly involves techniques too advanced to be presented here. The **Prime Number Theorem**, which was proved independently in 1896 by Hadamard and de la Vallée-Poussin, states that

$$\frac{\pi(x)}{x/\log_e x} \to 1 \quad \text{as} \quad x \to \infty,$$

where $\pi(x)$ is the number of primes $\leq x$.

There is a scarcity of results implying the existence of infinitely many primes of special forms. For example primes p and q are called **twin primes** if they differ by 2. Examples are 3 and 5, 5 and 7, 11 and 13, 41 and 43, and 1,000,000,009,649 and 1,000,000,009,651. When lists of large primes are compiled, twin primes appear to continue to pop up no matter how far out one goes, but the **twin prime conjecture**, which states that there are infinitely many of them, has never been proved.

The one positive result of this type is **Dirichlet's Theorem**, which says that if a and b are relatively prime, then the infinite arithmetic progression a, $a + b$, $a + 2b, \ldots$ contains infinitely many primes.

Factorization by Computer

The following BASIC program can be used to factor numbers of reasonable size. (Note its use of Theorem 2.3.) The program checks a number for divisibility by all integers up to its square root, since the computer doesn't know which divisors are prime. A more efficient (but longer) program would include a list of the smaller primes.

```
10 PRINT "THIS PROGRAM WILL TELL IF A NUMBER IS
PRIME, AND GIVE ITS SMALLEST FACTOR IF NOT. WHAT
IS YOUR NUMBER?"
20 INPUT N
30 FOR I = 2 TO INT(SQR(N))
40 IF N/I = INT(N/I) GOTO 80
50 NEXT I
60 PRINT N; "IS PRIME."
70 END
80 PRINT "THE SMALLEST DIVISOR OF"; N;
   "IS"; I; "."
90 END
```

Figure 2-1

Before the development of electronic computers mechanical devices were constructed to perform tedious number-theoretic computations. Figure 2-1 shows a photograph of a machine made by D.H. Lehmer of the University of California. Such devices played an important part in primality testing and other investigations before the birth of modern computers.

PROBLEMS FOR SECTION 2.5

 Use the Sieve of Eratosthenes to do each of the first four problems.

1. Find all primes between 50 and 100.
2. Find all primes between 100 and 150.
3. Find all primes between 1000 and 1025.
4. Find all primes between 1025 and 1050.

In the next two problems pretend the number P used in the first proof of Theorem 2.4 really exists.

5. Find (P, n) for $n = 11, 12, \ldots, 20$.
6. Find (P, n) for $n = 21, 22, \ldots, 30$.

B

7. Define a legitimate function $P(n)$ to take the place of the fictional (P, n).
8. Give a detailed proof of why the number $p_1 p_2 \cdots p_k + 1$ of Euclid's proof of Theorem 2.4 is not divisible by any of the p's.
9. How many primes are congruent to 0 (mod 4)? To 2 (mod 4)?
10. Show that if $(a, b) > 1$ then there exists at most one prime congruent to a (mod b).
11. Show that if p and q are twin primes and $p + q > 8$, then 12 divides $p + q$.
12. Show that if p, $p + 2$, and $p + 4$ are all prime, then $p = 3$.
13. Show that if $ab = c^2$, with a and b relatively prime positive integers, then $a = r^2$ and $b = s^2$ for some integers r and s.

C

14. Let S be a finite set of positive integers. Show, without using Theorem 2.4, that there exists $n > 1$ such that $(n, x) = 1$ for all x in S.
15. Disprove the statement: If p_i is the ith prime, then $p_1 p_2 \cdots p_k + 1$ is prime for all positive integers k.
16. Prove that there are infinitely many primes (such as 3, 7, 11, and 19) that are congruent to 3 (mod 4). (**Hint:** Suppose there are only $p_1, \ldots, p_k$. Show that $4p_1 p_2 \cdots p_k - 1$ must have a prime divisor $\equiv 3$ (mod 4). See problem 9.)
17. Prove that there are infinitely many primes congruent to 5 (mod 6).
18. Show that if k is any positive integer, then the numbers $(k + 1)! + j, j = 2, 3, \ldots, k + 1$, are all composite. (Thus there exist arbitrarily long sequences of consecutive composite numbers.)
19. Show by induction on n that for all positive integers n

$$1 + \frac{1}{2} + \frac{1}{3} + \cdots + \frac{1}{2^n} > \frac{n + 1}{2}.$$

Conclude that $1 + 1/2 + \cdots + 1/n$ can be made arbitrarily large.

The remaining problems are for those with some knowledge of infinite series from calculus.

20. Show that for k any positive integer

$$\left(1 + \frac{1}{p}\right)\left(1 + \frac{1}{p^2} + \frac{1}{p^4} + \cdots + \frac{1}{p^{2k}}\right) = 1 + \frac{1}{p} + \frac{1}{p^2} + \cdots + \frac{1}{p^{2k+1}}.$$

21. Let the ith prime be p_i, and let N be the set of all positive integers all of whose prime factors are $\leq p_k$. Use the last problem to show

$$\left(1 + \frac{1}{p_1}\right)\left(1 + \frac{1}{p_2}\right) \cdots \left(1 + \frac{1}{p_k}\right) \sum_{n \in N} \frac{1}{n^2} = \sum_{n \in N} \frac{1}{n}.$$

22. Conclude from the last problem that

$$\left(1 + \frac{1}{p_1}\right)\left(1 + \frac{1}{p_2}\right) \cdots \left(1 + \frac{1}{p_n}\right) \to \infty \quad \text{as} \quad n \to \infty.$$

23. Show that if $C > 0$, then $e^C > 1 + C$.
24. Use the last two problems to show that $1/p_1 + 1/p_2 + \cdots$ diverges. (**Hint:** Let $C = 1/p_i$.)
25. Let S be a finite set of positive integers, and let T be the set of all finite products of elements of S. Show that $\sum 1/n$ converges, where the sum is taken over T.
26. Show that if $f(x)$ is a polynomial of degree at least 2, then the set of values of $f(n)$, $n = 1, 2, \ldots$, does not contain all primes.
27. Show that if p_i is the ith prime, then

$$\left(1 - \frac{1}{p_1}\right)\left(1 - \frac{1}{p_2}\right) \cdots \left(1 - \frac{1}{p_n}\right) \to 0 \quad \text{as} \quad n \to \infty.$$

2.6   A Formula for $d(n)$

We now return to the evaluation of $d(n)$, which was sidetracked when we got into the much more interesting problem of prime factorization. What we know now makes the problem easy. Suppose

$$n = p_1^{k_1} p_2^{k_2} \cdots p_t^{k_t},$$

where the p's are distinct primes. If d is a positive divisor of n, then by the first part of Theorem 2.2

$$d = p_1^{j_1} p_2^{j_2} \cdots p_t^{j_t},$$

where $0 \leq j_i \leq k_i$ for $i = 1, 2, \ldots, t$. (We allow the j's to equal 0 to take care of primes not dividing d at all.)

For example, if $n = 63 = 3^2 7$, then the positive divisors of n are

$$3^0 7^0 \qquad 3^1 7^0 \qquad 3^2 7^0$$

$$3^0 7^1 \qquad 3^1 7^1 \qquad 3^2 7^1.$$

Returning to the general case, we note that the unique factorization theorem also tells us that each different choice for j_1 through j_t gives us a different d. Since there are $k_1 + 1$ possibilities for j_1, namely $0, 1, \ldots$, and $k_1, k_2 + 1$ possibilities for j_2, etc., we have the following theorem:

THEOREM   If $n = p_1^{k_1} p_2^{k_2} \cdots p_t^{k_t}$, where the p's are distinct primes, then

$$d(n) = (k_1 + 1)(k_2 + 1) \cdots (k_t + 1).$$

Example   Since $63 = 3^2 7$, we have $d(63) = (2 + 1)(1 + 1) = 6$. Likewise $d(120) = d(2^3 3^1 5^1) = (3 + 1)(1 + 1)(1 + 1) = 16$.

Writing Divisors in a Multiplication Table

When, before the theorem, we wrote out the positive divisors of 63, we found it convenient to organize them into a rectangular array. This array may have reminded the reader of a multiplication table; indeed, that is exactly what it was. Let us look at it again.

	1	3	9
1	1	3	9
7	7	21	63

We have written the positive divisors of 9 along the top and those of 7 along the left side. Each entry in the table is the product of a divisor of 9 and one of 7. Thus in this case

$$d(63) = d(9)d(7) = 3 \cdot 2 = 6.$$

This suggests that we might prove in general that $d(ab) = d(a)d(b)$ by showing that the divisors of ab are just all products of a divisor of a with one of b.

Let us suppose a and b have the positive divisors $d_1, d_2, \ldots, d_s$ and $e_1, e_2, \ldots, e_t$ respectively. Thus $d(a) = s$ and $d(b) = t$. We consider the array

$$
\begin{array}{cccc}
d_1 e_1 & d_2 e_1 & \cdots & d_s e_1 \\
d_1 e_2 & d_2 e_2 & \cdots & d_s e_2 \\
& & \vdots & \\
d_1 e_t & d_2 e_t & \cdots & d_s e_t.
\end{array}
$$

This array has st entries, so in order to show $d(ab) = st$ we must demonstrate three things:

1. every entry is a divisor of ab;
2. every positive divisor of ab appears in the array;
3. no two entries are equal.

Since (1) is easy to see we proceed to (2). Suppose c is a positive divisor of ab. Then there exists an integer k such that $ab = ck$. Consider the prime factorizations of both sides of this equation. By the Fundamental Theorem of Arithmetic we can match up the primes on both sides in a one-to-one fashion. In particular, some of the primes in the factorization of c match up with primes in the factorization of a, and the rest with primes in b. We see that c is a product of a divisor of a with one of b.

We now turn to condition (3). Suppose that two entries of the array are equal, say $d_i e_j = d_k e_l$. If we knew that the prime factors of d_i could only occur in d_k, and vice versa, we could conclude that $d_i = d_k$, and $e_j = e_l$ would follow. Unfortunately this need not be true. For example, if p is a prime dividing both a and b, then $p \cdot 1$ and $1 \cdot p$ will appear at different places in the array.

Evidently we need an additional hypothesis in order to prove (3), namely that no prime divides both a and b. If this is the case we easily see that $d_i e_j = d_k e_l$ implies that $d_i = d_k$ and $e_j = e_l$, since, for example, the primes dividing d_i cannot appear in e_l, a divisor of b, and so must all turn up in d_k. Since saying no prime divides both a and b is equivalent to saying $(a, b) = 1$, we have proved the following theorem:

THEOREM   If a and b are relatively prime, then $d(ab) = d(a)d(b)$.

This theorem could have been proved much more easily directly from the formula we developed for $d(n)$. We have proved somewhat more, however, namely the next theorem.

THEOREM 2.5   If a and b are relatively prime, the set of positive divisors of ab consists exactly of all products de, where d is a positive divisor of a and e is a positive divisor of b. Furthermore, these products are all distinct.

DEFINITION   multiplicative function
A numerical function $f(n)$ is said to be **multiplicative** if $f(ab) = f(a)f(b)$ whenever $(a, b) = 1$.

PROBLEMS FOR SECTION 2.6

In each of the first ten problems find $d(n)$ by means of the formula of this section.

1. $n = 75$ 2. $n = 45$ 3. $n = 30$
4. $n = 84$ 5. $n = 1000000$ 6. $n = 640$
7. $n = 900$ 8. $n = 2310$ 9. $n = 4961$
10. $n = 17640$

In each of the next four problems make a multiplication table of products of positive divisors of a times positive divisors of b. Are all the entries divisors of ab? Does every positive divisor of ab appear? Are the entries distinct?

11. $a = 8, b = 15$ 12. $a = 15, b = 66$
13. $a = 28, b = 21$ 14. $a = 21, b = 26$

15. Show that if $d \mid a$ and $e \mid b$, then $de \mid ab$.

In each of the next four problems show that the function f is multiplicative, that is, if $(a, b) = 1$, then $f(ab) = f(a)f(b)$. Is the hypothesis that $(a, b) = 1$ needed?

16. $f(n) = n$ 17. $f(n) = 1/n$
18. $f(n) = 1$ if n is odd and $f(n) = 0$ if n is even
19. $f(n) = (n, 12)$

True-False. In the next eight problems tell which statements are true and give counterexamples for the others. Assume f and g are multiplicative functions, that $g(n) = 0$ for no n, that C is a real number, and that $k > 0$. We define the functions $f + g$, fg, f^k, Cf, f/g, and $f \circ g$ at n to be $f(n) + g(n)$, $f(n)g(n)$, $(f(n))^k$, $Cf(n)$, $f(n)/g(n)$, and $f(g(n))$, respectively.

20. $f + g$ is multiplicative   21. fg is multiplicative
22. Cf is multiplicative   23. f^k is multiplicative
24. f/g is multiplicative   25. $f \circ g$ is multiplicative
26. $f(1) = 1$ 27. $g(1) = 1$

28. Find the smallest n such that $d(n) = 35$.
29. Find x such that $d(10x) = 10$.
30. Find x such that $d(10x) = x$.
31. Show that if p is prime, then there exists x such that $d(px) = x$.
32. Show that if p is prime, then there exists x such that $d(p^2 x) = x$.
33. Show that if p is prime, then there exists x such that $d(2px) = x$.

34. Prove that $d(n)$ is multiplicative by means of the formula of this section.
35. Prove that if $(a, b) > 1$ then $d(ab) < d(a)d(b)$.
36. Prove that $d(n)$ is odd if and only if n is a perfect square.
37. Show that if $m > 1$, then there exist infinitely many integers n such that $d(n) = m$.
38. Show that if $n = k^m$, then $d(n) \equiv 1 \pmod{m}$.
39. Show that if $d(n)$ is prime, then n is a power of a prime.
40. Show that if $n > 0$ then as d runs through the positive divisors of n, so does n/d.
41. Show that if $d \mid a$, $e \mid b$, and $(a, b) = 1$, then $(d, e) = 1$.
42. Show that if $n > 0$, then there exists x such that $d(nx) = n$.
43. Show that if $n > 0$, the number of solutions to $d(nx) = n$ is 1 if and only if $n = 1$, 4, or a prime, is finite if n is the product of distinct primes, and otherwise is infinite.
44. Find the smallest $n > 0$ such that $d(nx) = x$ has no solution.
45. Show that $d(d(n)) = 2$ if and only if $n = p^{q-1}$, where p and q are primes.

Pierre de Fermat

1608–1665

The great French scientific amateur Pierre de Fermat led a quiet life practicing law in Toulouse, and producing high quality work in number theory and other areas of mathematics as a hobby. He published almost nothing, revealing most of his results in his extensive correspondence with friends, and generally kept his proofs to himself. "Fermat's Theorem" first appeared in a letter to Frenicle de Bessy dated October 18, 1640.

Fermat also worked with differential calculus, finding a way of calculating maxima and minima. He reasoned that if $f(x)$ had an extremum at x, then changing x slightly would not have much effect on $f(x)$. For example, if $f(x) = x^2 + x$, then, letting e represent the change in x, he set $f(x + e) = f(x)$. This gives

$$x^2 + 2xe + e^2 + x + e = x^2 + x,$$

or, after dividing by e,

PIERRE DE FERMAT

$$2x + e + 1 = 0.$$

Now setting $e = 0$ gives $x = -\frac{1}{2}$, which is the x minimizing $f(x)$.

Calculus students will recognize that this amounts to calculating the derivative of f from the definition and setting it equal to 0. The reason most people do not credit Fermat with being the founder of differential calculus (as was claimed by his countrymen Lagrange, Laplace, and Fourier) is that, although Fermat could solve individual problems such as the one given, he did not develop general differentiation rules as did Newton and Leibniz. Nonetheless in a letter that came to light only in 1934 Newton did acknowledge that he got from Fermat's method of constructing tangents the hint that led to the development of the differential calculus.

Chapter 3

Numerical Functions

Functions like $d(n)$, the number of positive divisors of n, defined on the positive integers, are variously called numerical, arithmetic, or number-theoretic functions. They are treated in this chapter.

3.1 The Sum of the Divisors

DEFINITION $\sigma(n)$

Let n be a positive integer. We define $\sigma(n)$ to be the sum of the positive divisors of n.

Examples If $n = 7$ we have $\sigma(n) = 1 + 7 = 8$; for $n = 9$ we have $\sigma(n) = 1 + 3 + 9 = 13$; and for $n = 63$ we have

$$\sigma(n) = \quad 1 + \; 3 + \; 9$$
$$+ \; 7 + 21 + 63 = 104.$$

The Multiplication Table Again

The astute reader may have noticed that $\sigma(63) = 104 = \sigma(7)\sigma(9)$, and that in fact in adding up the positive divisors of 63 we have again created a multiplication table of the divisors of 7 times those of 9. We are seeing a repetition of the ideas leading to our proof that $d(n)$ was multiplicative. Indeed, Theorem 2.5 of the last section may be used to prove $\sigma(n)$ multiplicative just as it was used to prove $d(n)$ multiplicative. The reader may want to try to provide his or her own proof for the next theorem before reading the one given.

THEOREM 3.1   The function $\sigma(n)$ is multiplicative.

Proof   Suppose a and b are relatively prime positive integers. Let the positive divisors of a be $d_1, d_2, \ldots, d_s$, and let those of b be $e_1, e_2, \ldots, e_t$. Then

$$
\begin{aligned}
\sigma(a)\sigma(b) &= (d_1 + d_2 + \cdots + d_s)(e_1 + \cdots + e_t) \\
&= \quad d_1 e_1 + d_2 e_1 + \cdots + d_s e_1 \\
&\quad + d_1 e_2 + d_2 e_2 + \cdots + d_s e_2 \\
&\qquad\qquad \vdots \\
&\quad + d_1 e_t + d_2 e_t + \cdots + d_s e_t.
\end{aligned}
$$

But by Theorem 2.5 the numbers in this sum are exactly the positive divisors of ab. Thus $\sigma(a)\sigma(b) = \sigma(ab)$.

The Summation Notation

There is a notation for sums that avoids the sprawl (and possible ambiguity) of the . . . 's used above. The Greek letter $\sum$ (capital sigma) signals the sum of terms, each of which depends on an integer that varies between two limits. For example the sum

$$d_1 + d_2 + \cdots + d_s$$

of interest in the last theorem would be expressed as

$$\sum_{i=1}^{s} d_i.$$

In general, if f is some function of an integral variable and $m \leq n$, by

$$\sum_{i=m}^{n} f(i)$$

we mean

$$f(m) + f(m + 1) + f(m + 2) + \cdots + f(n).$$

The variable i is called the **index of summation** and is similar to the variable of integration in a definite integral in that it really does not matter what letter is used. For example,

$$\sum_{i=1}^{5} i \quad \text{and} \quad \sum_{j=1}^{5} j$$

both mean the same thing (namely, 15), and any other letter could be used in place of i or j so long as it had no previous meaning.

In number theory it is common to extend the summation notation to situations in which the index of summation does not run through a set of consecutive integers. A description of the values the index is allowed to assume is simply written under the $\sum$. For example,

$$\sum_{\substack{p \text{ prime} \\ p < 7}} p^2 = 4 + 9 + 25 = 38, \qquad \text{and} \qquad \sigma(n) = \sum_{\substack{d|n \\ d > 0}} d$$

To interpret the last summation correctly the reader must realize that d and not n is the index of summation. One clue to this is that the n appears outside the summation (in $\sigma(n)$), something an index of summation never does.

When summing over the divisors of a number we generally wish to consider those that are positive, so in order to avoid always writing $d > 0$ under the sigma we establish the following convention:

CONVENTION   When a summation is over the divisors of a number, these divisors are restricted to be positive.

There are various rules for manipulating summations that follow from the usual rules of algebra. For example,

$$\sum_{i=1}^{n} Kf(i) = K \sum_{i=1}^{n} f(i),$$

because

$$Kf(1) + Kf(2) + \cdots + Kf(n) = K(f(1) + \cdots + f(n))$$

according to the distributive law. Similar rules will be found in the exercises and can be proved by reverting to the ... notation.

One way to express the number of elements in a set is to sum 1 over that set. For example,

$$d(n) = \sum_{d|n} 1.$$

Evaluating Multiplicative Functions

Knowing that a function is multiplicative is valuable because it reduces the problem of determining a formula for it to evaluating the function at a power of a prime. Suppose, for example, $n = p_1^{k_1} p_2^{k_2} \cdots p_m^{k_m}$, where the p's are distinct primes, and suppose we know f is a multiplicative function. Then

$$f(n) = f(p_1^{k_1}p_2^{k_2} \cdots p_m^{k_m}) = f(p_1^{k_1})f(p_2^{k_2} \cdots p_m^{k_m})$$
$$= f(p_1^{k_1})f(p_2^{k_2})f(p_3^{k_3} \cdots p_m^{k_m})$$
$$= \qquad \cdots$$
$$= f(p_1^{k_1})f(p_2^{k_2}) \cdots f(p_m^{k_m}).$$

Evaluating f at a prime power is generally simpler than at an arbitrary integer. For example,

$$\sigma(p^k) = 1 + p + p^2 + \cdots + p^k.$$

This is a geometric progression, and by the formula we developed in Section 1.6 its sum is

$$\frac{p^{k+1} - 1}{p - 1}.$$

We are now in a position to give a formula for the function $\sigma(n)$, but the notation we have would again involve an excessive use of ellipsis (three dots). We need a compact symbolism for products analogous to the sigma notation for sums.

The Product Notation

The symbol $\prod$ (capital Greek letter pi) is used to denote a product. Thus

$$\prod_{i=1}^{n} f(i) = f(1)f(2) \cdots f(n)$$

and, more generally, if $S(i)$ is a statement about i, by

$$\prod_{S(i)} f(i)$$

we mean the product of the numbers $f(i)$ over all values of i for which $S(i)$ is true.

Examples

$$\prod_{i=1}^{3} i^2 = 1 \cdot 4 \cdot 9 = 36,$$

$$\prod_{\substack{p \text{ prime} \\ p < 5}} \frac{1}{p} = \frac{1}{2} \cdot \frac{1}{3} = \frac{1}{6},$$

and

$$\prod_{i=1}^{n} i = n!.$$

THEOREM 3.2   If $n = p_1^{k_1} p_2^{k_2} \cdots p_m^{k_m}$, where the p's are distinct primes, then

$$\sigma(n) = \prod_{i=1}^{m} \frac{p_i^{k_i + 1} - 1}{p_i - 1}.$$

Examples   If $n = 63 = 7 \cdot 3^2$, then

$$\sigma(n) = \frac{7^2 - 1}{7 - 1} \frac{3^3 - 1}{3 - 1} = \left(\frac{48}{6}\right)\left(\frac{26}{2}\right) = 8 \cdot 13 = 104.$$

Likewise

$$\sigma(1000000) = \sigma(2^6 5^6) = \frac{2^7 - 1}{2 - 1} \frac{5^7 - 1}{5 - 1} = \left(\frac{127}{1}\right)\left(\frac{78124}{4}\right) = 2480437.$$

PROBLEMS FOR SECTION 3.1

 In the first 10 problems compute $\sigma(n)$.

1. $n = 20$ 2. $n = 30$ 3. $n = 100$ 4. $n = 81$
5. $n = 101$ 6. $n = 40$ 7. $n = 667$ 8. $n = 1000$
9. $n = 272$ 10. $n = 1331$

Define $s_2(n)$ to be $\sum_{d|n} d^2$.

Compute $s_2(n)$ in each of the next 4 problems.

11. $n = 7$ 12. $n = 9$ 13. $n = 63$ 14. $n = 100$

Define $s_{-1}(n)$ to be $\sum_{d|n} (1/d)$.

Compute $s_{-1}(n)$ in each of the next 4 problems.

15. $n = 7$ 16. $n = 9$ 17. $n = 63$ 18. $n = 28$

In the next 12 problems compute each sum or product.

19. $\displaystyle\sum_{i=1}^{5} i^2$ 20. $\displaystyle\sum_{i=3}^{6} \frac{1}{i}$ 21. $\displaystyle\sum_{d|12} (d + 0.5)$

22. $\displaystyle\sum_{d|18} \frac{6}{d}$ 23. $\displaystyle\sum_{d|128} 1$ 24. $\displaystyle\sum_{i=100}^{200} 1$

25. $\displaystyle\sum_{\substack{|p-10|<5 \\ p \text{ prime}}} p$ 　　　　26. $\displaystyle\sum_{\substack{p \text{ prime} \\ p-1 \text{ square} \\ p<20}} p$ 　　　　27. $\displaystyle\prod_{i=3}^{5} \frac{i}{i+1}$

28. $\displaystyle\prod_{i=-2}^{1} i$ 　　　　　　29. $\displaystyle\prod_{d|10} (d+1)$ 　　　　　30. $\displaystyle\prod_{\substack{p \text{ prime} \\ p|24}} p^2$

B *Prove the formula given in each of the next 4 problems. All sums and products run from $i = 1$ to n.*

31. $\displaystyle\sum (f(i) + g(i)) = \sum f(i) + \sum g(i)$ 　　　　32. $\displaystyle\sum K = nK$

33. $\displaystyle\prod Kf(i) = K^n \prod f(i)$ 　　　　　　　　34. $\displaystyle\prod (f(i))^K = \left(\prod f(i)\right)^K$

35. Show by induction on n that

$$\sum_{i=1}^{n} Kf(i) = K \sum_{i=1}^{n} f(i),$$

using only that $K(a + b) = Ka + Kb$.

36. Suppose $m \le n$. What is

$$\sum_{i=m}^{n} 1?$$

37. Give a noninduction proof of the formula for the sum of a geometric progression of Section 1.6 by letting the sum be S and showing that $RS - S = A(R^n - 1)$.

C 38. Show that the function s_2 defined above is multiplicative.
39. Show that the function s_{-1} defined above is multiplicative.
40. Show that if p is prime then

$$s_2(p^k) = \frac{p^{2(k+1)} - 1}{p^2 - 1}.$$

41. Show that if p is prime then

$$s_{-1}(p^k) = \frac{p^{k+1} - 1}{p^k(p - 1)}.$$

42. Use Problem 40 of Section 2.6 to show that $n \cdot s_{-1}(n) = \sigma(n)$ for all n.
43. Use Problem 41 of this section to show that $n \cdot s_{-1}(n) = \sigma(n)$ for all n.
44. Show that if $\sigma(n)$ is odd then $n = j^2 2^k$, where j is odd.

45. Show that

$$\left(\sum_{i=1}^{m} f(i)\right)\left(\sum_{j=1}^{n} g(j)\right) = \sum_{i=1}^{m}\left(\sum_{j=1}^{n} f(i)g(j)\right)$$

$$= \sum_{j=1}^{n}\left(\sum_{i=1}^{m} f(i)g(j)\right).$$

46. Prove by induction that if $p_1, p_2, \ldots, p_k$ are distinct primes and if f is multiplicative, then

$$f(p_1^{m_1}p_2^{m_2}\cdots p_k^{m_k}) = f(p_1^{m_1})f(p_2^{m_2})\cdots f(p_k^{m_k}).$$

3.2   Multiplicative Functions

Manufacturing Multiplicative Functions

In the last two sections we proved first $d(n)$ and then $\sigma(n)$ to be multiplicative, using almost the same proof. In the last problem set the functions $s_2(n)$ and $s_{-1}(n)$ were defined to be respectively the sum of the squares and reciprocals of the positive divisors of n, and these both turned out to be multiplicative also. Let us see how far this method of proof can be carried.

Let us suppose that f is some arbitrary numerical function, and define the function F by

$$F(n) = \sum_{d|n} f(d).$$

As before, let us assume that a and b are relatively prime, with positive divisors $d_1, d_2, \ldots, d_s$ and $e_1, e_2, \ldots, e_t$, respectively. Then

$$F(a)F(b) = (f(d_1) + \cdots + f(d_s))(f(e_1) + \cdots + f(e_t))$$

$$= \quad f(d_1)f(e_1) + \cdots + f(d_s)f(e_1)$$
$$+ f(d_1)f(e_2) + \cdots + f(d_s)f(e_2)$$
$$\vdots$$
$$+ f(d_1)f(e_t) + \cdots + f(d_s)f(e_t).$$

On the other hand, by Theorem 2.5 $F(ab)$ is

$$f(d_1e_1) + \cdots + f(d_se_1)$$
$$+ f(d_1e_2) + \cdots + f(d_se_2)$$
$$\vdots$$
$$+ f(d_1e_t) + \cdots + f(d_se_t).$$

We see that in order for these two expressions to be equal, it is enough that $f(d)f(e) = f(de)$ whenever $d|a$ and $e|b$. Since in such circumstances d and e will be relatively prime if a and b are, it suffices that the function f be multiplicative in order that F is.

THEOREM 3.3   Suppose f is a multiplicative function, and define F by

$$F(n) = \sum_{d|n} f(d).$$

Then F is multiplicative also.

Examples   Since the function defined by $f(n) = n$ for all n is easily seen to be multiplicative, then so is

$$\sigma(n) = \sum_{d|n} f(d) = \sum_{d|n} d.$$

In fact if we define $f(n)$ to be n^k, we easily see f is multiplicative. Thus so is

$$s_k(n) = \sum_{d|n} f(d) = \sum_{d|n} d^k.$$

Thus at one swoop we have shown $\sigma(n)$ (taking $k = 1$), $d(n)$ (taking $k = 0$), and the functions of the problems of the last section (with $k = 2$ and -1) to be multiplicative.

We now have a general scheme for inventing and finding formulas for multiplicative functions of the form

$$F(n) = \sum_{d|n} f(d).$$

We start with some simple function f that we prove to be multiplicative. Then by the last theorem F is also multiplicative. To find a formula for F (or for any multiplicative function) it suffices to evaluate F at prime powers, just as in our derivation of the formula for $\sigma(n)$ in Section 3.1.

As an example we consider the function $g(n)$ defined to be 0 if n is even and 1 if n is odd. This is easily seen to be multiplicative since $g(ab)$ and $g(a)g(b)$ are both 1 if and only if a and b are both odd. Now let

$$G(n) = \sum_{d|n} g(d)$$

Then G is multiplicative by the last theorem. We evaluate G at prime powers. The prime 2 is a special case. We have

$$G(2^k) = g(1) + g(2) + \cdots + g(2^k) = 1.$$

On the other hand, if p is an odd prime then

$$G(p^k) = g(1) + g(p) + \cdots + g(p^k) = k + 1.$$

We see that if $n = 2^k p_1^{k_1} p_2^{k_2} \cdots p_s^{k_s}$, where the p's are distinct odd primes, then $G(n) = (k_1 + 1)(k_2 + 1) \cdots (k_s + 1)$. This is similar to the formula for $d(n)$. In fact $G(n)$ is the number of *odd* divisors of n.

Why Study Number Theory?

Perhaps the reader is wondering what good a function like $\sigma(n)$ is. Who cares that we have found a formula for the sum of the positive divisors of n? This is a question that each person must answer for himself or herself, since number theory, like mountain climbing, is an endeavor practiced mainly just for its own sake. Among the reasons some people might care about it are:

Accomplishment Solving a mathematical problem may give the same pleasant feeling of success as finishing a difficult crossword puzzle or winning a set of tennis.

Understanding Knowing why a number is divisible by 9 if and only if the sum of its digits is (and thus the justification of the old check of "casting out 9's") may bring satisfaction.

Skill A person might take pride in being able to play the guitar or water ski, or, just as well, count the divisors of a million in a few seconds or compute the greatest common divisor of two large numbers.

Beauty Many find beauty in number theory, either in the structure of the integers themselves (as an astronomer might find beauty in the laws of the universe), or in an ingenious or elegant proof.

We do not want to leave the impression that there are no applications at all for number theory. The RSA method of public key cryptography, which is treated in Section 4.5, involves many number-theoretic concepts, and has gained a great deal of attention in recent years. Its security depends on the difficulty of factoring large numbers, a problem that has been studied for centuries, long before its connection to anything useful. This is similar to many episodes in science in which phenomena studied purely for their theoretical interest later turned out to have immense practical consequences (such as atomic energy).

The study of numbers for their own sake is by no means new. The Greeks were interested in numbers like 6, which is the sum of its positive divisors other than itself; $6 = 1 + 2 + 3$. If n is any number with this property, then $n = \sigma(n) - n$, since by definition $\sigma(n)$ adds in the divisor n itself.

DEFINITION   perfect number
We say n is a **perfect number** in case $\sigma(n) = 2n$.

A Table for $\sigma(n)$

To look for perfect numbers other than 6 we will make a table of the first 30 values of $\sigma(n)$ as follows:

n	$\sigma(n)$	n	$\sigma(n)$	n	$\sigma(n)$
1	1	11	12	21	32
2	3	12	28	22	36
3	4	13	14	23	24
4	7	14	24	24	60
5	6	15	24	25	31
6	12	16	31	26	42
7	8	17	18	27	40
8	15	18	39	28	56
9	13	19	20	29	30
10	18	20	42	30	72

Construction of the table was simplified by using a few properties of $\sigma(n)$. The primes were easily filled in, since $\sigma(p) = p + 1$. Prime powers follow the rule that $\sigma(p^k) = \sigma(p^{k-1}) + p^k$. Thus $\sigma(4) = \sigma(2) + 4$, $\sigma(8) = \sigma(4) + 8$, etc. The remaining values were then determined by the fact that $\sigma(n)$ is multiplicative. For example $\sigma(18) = \sigma(2)\sigma(9) = 39$. (But $\sigma(3)\sigma(6) = 48$. What's wrong?)

Interest in perfect numbers precedes Euclid, and many early writers made note of them. Saint Augustine said that God created the world in six days rather than all at once because "The perfection of the work is signified by the perfect number 6." Others asserted that there are infinitely many perfect numbers; that, indeed, there is exactly one between 1 and 10, another between 10 and 100, and another between 100 and 1000, etc.; that all perfect numbers are even; and that the perfect numbers alternately end in the digits 6 and 8. None of these assertions has ever been proved, and some are known to be false. We will continue the study of perfect numbers in the next section.

DEFINITIONS   abundant and deficient numbers
A number is said to be **abundant** if the sum of the positive divisors of the number other than itself exceeds the number (so $\sigma(n) > 2n$), and **deficient** if this sum is less than the number (so $\sigma(n) < 2n$).

Examples   Since $\sigma(12) = 28 > 2 \cdot 12$, the number 12 is abundant. Since $\sigma(11) = 12 < 2 \cdot 11$, the number 11 is deficient. Clearly each positive integer is either perfect, abundant, or deficient.

DEFINITION amicable numbers

We say a pair of numbers a and b is **amicable** if the sum of the positive divisors of a less than a equals b, and the sum of the positive divisors of b less than b equals a. Another way to say this is that

$$\sigma(a) - a = b \quad \text{and} \quad \sigma(b) - b = a.$$

Example The smallest pair of amicable numbers is 220 and 284. Here $220 = 2^2 5 \cdot 11$, and

$$\sigma(220) - 220 = \sigma(2^2)\sigma(5)\sigma(11) - 220 = 7 \cdot 6 \cdot 12 - 220 = 284.$$

Likewise $284 = 2^2 71$, and

$$\sigma(284) - 284 = \sigma(2^2)\sigma(71) - 284 = 7 \cdot 72 - 284 = 220.$$

Amicable pairs have long been considered important in numerology (which has the same relation to number theory as astrology has to astronomy). The following is from the writings of the Arab scholar Ibn Khaldun (1332–1406):

> Let us mention that the practice of the art of talismans has also made us recognize the marvelous virtues of amicable (or sympathetic) numbers. These numbers are 220 and 284. One calls them amicable because the aliquot parts of one when added give a sum equal to the other. Persons who occupy themselves with talismans assure that these numbers have a particular influence in establishing union and friendship between two individuals. One prepares a horoscope theme for each individual, the first under the sign of Venus while this planet is in its house or in its exaltation and while it presents in regard to the moon an aspect of love and benevolence. In the second theme the ascendant should be in the seventh sign. On each one of these themes one inscribes one of the numbers just indicated, but giving the strongest number to the person whose friendship one wishes to gain, the beloved person. I don't know if by the strongest number one wishes to designate the greatest one or the one which has the greatest number of aliquot parts. There results a bond so close between the two persons that they cannot be separated. The author of the Ghaia and other great masters in this art declare that they have seen this confirmed by experience.

The pair 220, 284 was the only one known to the ancients, but later many pairs of amicable numbers were found. Now about 400 pairs are known. In 1866 a sixteen-year-old Italian boy, Nicolo Paganini, found the fairly small pair 1184, 1210, which had been overlooked by mathematicians up to that time.

At a meeting of the American Mathematical Society in San Francisco in 1981 Hilton Chen and Dale Woods of Northeast Missouri State University announced two previously unpublished pairs of amicable numbers, the larger of which was

$a = 217412690408758900835667725725679359793688363643843$
$b = 222617239908155568290127696866529750576199429564477.$

PROBLEMS FOR SECTION 3.2

1. Extend the table of $\sigma(n)$ to run from 31 to 50.
2. Extend the table of $\sigma(n)$ to run from 51 to 70.
3. Which of the numbers from 1 to 15 are abundant?
4. Which of the numbers from 16 to 30 are deficient?
5. Let $G(n)$ be the function defined in this section. Compute $G(n)$ for n from 1 to 10.
6. Compute $G(n)$ for n from 11 to 20.

We define the function $M(n)$ to be 1 if n is 1, 0 if $p^2|n$ for any prime p, and $(-1)^k$ if $n = p_1 p_2 \cdots p_k$, where the p's are distinct primes.

7. Compute $M(n)$ for $n = 1$ to 10.
8. Compute $M(n)$ for $n = 11$ to 20.
9. Check that 1184 and 1210 really are amicable.

In problems 10 and 11 assume f is a multiplicative function with the values given. Complete the table so far as you can be sure of the values.

10.

n	2	3	4	5	6	10
$f(n)$	3	4				8

11.

n	1	2	3	4	6	12
$f(n)$		4			5	

12. Prove that M (defined above) is multiplicative.
13. Show that if f and g are multiplicative functions, then so is fg, where $(fg)(n) = f(n)g(n)$.
14. Let $F(n) = \sum_{d|n} M(d)$. Compute $F(n)$ for $n = 1, 3, 4, 6$.
15. Let F be as in the last problem. Suppose p is prime and $k > 0$. Show $F(p^k) = 0$.
16. Let F be as in the last problem. Show $F(n) = 1$ if $n = 1$ and 0 otherwise.
17. Let $F(n) = \sum_{e|n} d(e)$. Compute $F(n)$ for $n = 1, 3, 4,$ and 6.
18. Let F be as in the last problem. Suppose p is prime. Show $F(p^k) = (k + 1)(k + 2)/2$.
19. Let F be as in the last problem and let $n = p_1^{m_1} p_2^{m_2} \cdots p_k^{m_k}$. Give a formula for $F(n)$.
20. Let $F(n) = \sum_{e|n} M(e)d(e)$. Give a formula for F.

21. Let $F(n) = \sum_{d|n} M(d)\sigma(d)$. Give a formula for F.

22. Show that all prime powers are deficient.

23. Show that a and b form an amicable pair if and only if $\sigma(a) = \sigma(b) = a + b$.

24. One of an amicable pair is 17296. What is the other?

$\boxed{\textbf{C}}$ 25. If $F(n) = \sum_{d|n} f(d)$ write $F = f'$. Show that $M'''(n) = d(n)$ for all n.

26. Let $m(n) = 0$ if $p^3 | n$, p prime. Otherwise let $m(n) = (-2)^k$, where k is the number of distinct primes p such that $p \| n$. Show that $m' = M$ in the notation of the last problem.

27. Suppose $f' = g'$, in the notation of problem 25. Show that $f = g$. (**Hint:** Use Induction II.)

28. Prove that if f and g are multiplicative, then so is

$$F(n) = \sum_{d|n} f(d)g\left(\frac{n}{d}\right)$$

29. Show that if f is multiplicative and if

$$F(n) = \sum_{d|n} f(d)$$

for all n, then

$$f(n) = \sum_{d|n} M(d)F\left(\frac{n}{d}\right)$$

for all n.

30. Let $F(n) = d(n) \Big/ \sum_{e|n} 1/e$. Show that $F(n) = nd(n)/\sigma(n)$ for all positive integers n.

3.3   Perfect Numbers

Looking for a Pattern

So far we have found only two perfect numbers: 6 and 28. Let us compute $\sigma(n)$ for each of these to try to see why it equals $2n$.

$$\sigma(6) = \sigma(\overbrace{2 \cdot 3}) = \sigma(2)\sigma(3) = 3 \cdot 4$$

$$\sigma(28) = \sigma(\overbrace{4 \cdot 7}) = \sigma(4)\sigma(7) = 7 \cdot 8$$

Each of the perfect numbers we know consists of a power of 2 times a prime. As the arrows indicate, when we apply σ to each, the power of 2 turns into the prime and the prime turns into twice the power of 2.

What must be the relation between the prime and power of 2 for this to work? Suppose $n = 2^r p$, where p is prime. To get the same pattern we need

$$\sigma(2^r) = 2^{r+1} - 1 = p$$

and

$$\sigma(p) = p + 1 = 2 \cdot 2^r = 2^{r+1}.$$

Notice that both these equations amount to the same thing, namely $p = 2^{r+1} - 1$. It looks as if whenever we can find a prime p that is one less than a power of 2, say $p = 2^k - 1$, then $2^{k-1}p$ will be a perfect number. (Here k is the $r + 1$ of the displayed equations.) This is true, but we will write out a formal proof just to make sure.

THEOREM If $2^k - 1$ is prime, then $2^{k-1}(2^k - 1)$ is perfect.

Proof Since $2^k - 1$ is an odd prime, it is relatively prime to 2^{k-1}. Thus by the multiplicativity of σ we have

$$\sigma(2^{k-1}(2^k - 1)) = \sigma(2^{k-1})\sigma(2^k - 1).$$

Now by Theorem 3.2

$$\sigma(2^{k-1}) = \frac{2^k - 1}{2 - 1} = 2^k - 1,$$

while $\sigma(2^k - 1) = 2^k - 1 + 1 = 2^k$.

We see that if $n = 2^{k-1}(2^k - 1)$, then

$$\sigma(n) = (2^k - 1)2^k = 2n.$$

Finding More Perfect Numbers

We now have a scheme for finding perfect numbers, being able to construct one whenever we find a prime that is one less than a power of 2.

k	$2^k - 1$	prime?	perfect number
1	1	no	
2	3	yes	$2 \cdot 3 = 6$
3	7	yes	$2^2 7 = 28$
4	15	no	
5	31	yes	$2^4 31 = 496$
6	63	no	

The reader should check that 496 really is perfect. Clearly the determination of when $2^k - 1$ is prime is an important question in the study of perfect numbers. We defer this until the next section, however, and turn instead to the question of whether or not we are on the track of all perfect numbers.

Are All Perfect Numbers $2^{k-1}(2^k - 1)$?

That each prime of the form $2^k - 1$ gives rise to a perfect number was known to Euclid, and the theorem we just proved appears in his *Elements*. This left the question "Are there any perfect numbers not of Euclid's form?" It took about 2000 years until progress was made on this question, and even then a complete answer was not given.

THEOREM (Euler)   Every even perfect number is of the form

$$2^{k-1}(2^k - 1), \qquad \text{where } 2^k - 1 \text{ is prime.}$$

Proof   Suppose n is an even perfect number. Then we can write

$$n = 2^r q, \quad r > 0,$$

where q is some odd integer. Using the multiplicativity of σ and the fact that q is odd, we have

$$\sigma(n) = \sigma(2^r)\sigma(q) = (2^{r+1} - 1)\sigma(q).$$

Note that we know nothing about q other than that it is odd, so we cannot evaluate $\sigma(q)$. From the fact that n is perfect, however, we have

$$\sigma(n) = 2 \cdot 2^r q = 2^{r+1} q.$$

Combining this with the previous equation and dividing through by $2^{r+1}\sigma(q)$ give

$$\frac{2^{r+1} - 1}{2^{r+1}} = \frac{q}{\sigma(q)}.$$

This equation provides the key to the proof of this theorem. Notice that the fraction on the left is close to 1, since the numerator is only one less than the denominator. On the other hand, we would expect the fraction on the right to be smaller, since its denominator $\sigma(q)$ is a sum including q, 1, and whatever other divisors q has. In fact we will show that q must be prime, so that it has no more divisors.

Of course just because two fractions are equal does not mean that the same goes for their numerators and denominators; the fractions might not be in lowest terms. In the last equation the fraction on the left is in lowest terms, since the numerator is odd, but it is not clear that the fraction on the right is. Let us suppose

$$(q, \sigma(q)) = d.$$

Then

$$q = d(2^{r+1} - 1) = d \cdot 2^{r+1} - d$$

and

$$\sigma(q) = d \cdot 2^{r+1} = q + d.$$

But $2^{r+1} - 1 > 1$ so d is a divisor of q other than q itself. If $d > 1$ then by definition $\sigma(q)$ is at least $q + d + 1$, which contradicts the last equation. We conclude that $d = 1$.

But setting $d = 1$ in the last equation gives $\sigma(q) = q + 1$. This clearly implies that q is prime. The previous equation then says that

$$q = 2^{r+1} - 1.$$

Setting $k = r + 1$ now yields the statement of the theorem.

Odd Perfect Numbers

Euler's theorem only covers even perfect numbers; are there any odd ones? Nobody knows. No one has ever found an odd perfect number, but neither has anyone ever shown that none exist, and there might even be infinitely many of them. It has been shown that if n is an odd perfect number, then

$$n > 100,000,000,000,000,000,000,000,000,000,000,000,000,000,000,000.$$

PROBLEMS FOR SECTION 3.3

1. Find the next perfect number after 496.
2. Extend the table given in this section to run from $k = 7$ to 10.

3. An early author stated that if k is odd then $2^k - 1$ is prime. Show that this is false.
4. An early author stated that if k is prime then $2^k - 1$ is prime. Show that this is false.

In the next five problems (a) give the prime power factorization of $2^k - 1$; (b) list all numbers $2^d - 1$, where d divides k, $1 < d < k$.

5. $k = 4$ 6. $k = 6$ 7. $k = 8$ 8. $k = 9$ 9. $k = 10$

10. For which numbers k is $2^k + 1$ prime, $k = 1, 2, \ldots, 10$?

In the next six problems define r by $2^r \| k$ if k is even. If k is odd let $r = 0$. For each k given, (a) write the prime power factorization of $2^k + 1$; (b) compute r; (c) compute $2^{2^r} + 1$.

11. $k = 3$ 12. $k = 5$ 13. $k = 6$
14. $k = 7$ 15. $k = 9$ 16. $k = 10$

17. Where does the proof of Euler's theorem fail if n is not even?
18. The Greeks knew of only five perfect numbers, the largest of which was 33550336. Of course this is of the Euclid type, $2^{k-1}p$. What are k and p?
19. It has been shown that $2^k - 1$ is prime for $k = 11213$. How many digits does this number have? (**Hint:** $\log_{10}2 = 0.301029996$)
20. Show that if $k > 0$ then 2^k, $2^k + 1$, and $2^k - 1$ all have the same number of digits.
21. Show by induction on n that if m and n are positive integers then $2^m - 1$ divides $2^{mn} - 1$.
22. Use the formula for the sum of a geometric progression from Section 1.6 to show that

$$1 + 2^m + 2^{2m} + \cdots + 2^{(n-1)m} = \frac{2^{mn} - 1}{2^m - 1}.$$

23. Show by use of congruence properties that

$$2^{mn} \equiv 1 \pmod{2^m - 1} \quad \text{for all positive integers } m \text{ and } n.$$

24. Show that any of the last 3 problems implies that for $2^k - 1$ to be prime, k must be prime.
25. Recall the function $s_{-1}(n)$ defined in the problems for Section 3.1. Show that n is perfect if and only if $s_{-1}(n) = 2$.
26. Show that if n is an odd perfect number, then $n = pa^2$, where p is prime. (**Hint:** Consider the equation $\sigma(n) = 2n$. When is $\sigma(p^k)$ odd?)
27. We could replace addition by multiplication in the definition of perfect numbers and ask when n is the product of its positive divisors other than

itself. Show that this happens when $n = p^3$, p prime, or $n = pq$, p and q distinct primes.
28. Show that if $n > 1$, then n is the product of its positive divisors other than itself only in the cases mentioned in the previous problem.
29. According to Book IX of Euclid's *Elements* "If as many numbers as we please, beginning from unity, be set out continuously in double proportion until the sum of all becomes prime, and if the sum is multiplied by the last, the product will be perfect." Explain this.
30. Show that no perfect number is a square.
31. Show that if p is as in problem 26, then $p \equiv 1$ (mod 4).

3.4 Mersenne and Fermat Numbers

DEFINITION Mersenne numbers

If k is a positive integer we call the number $2^k - 1$ a **Mersenne number**, and denote it by M_k.

Which Mersenne Numbers Are Prime?

Of course we are interested in this question because we have found that there is a one-to-one correspondence between Mersenne primes and even perfect numbers. Let us carry the table of the last section further.

k	$M_k = 2^k - 1$	prime?
1	1	no
2★	3	yes
3★	7	yes
4	$15 = 3 \cdot 5$	no
5★	31	yes
6	$63 = 9 \cdot 7$	no
7★	127	yes
8	$255 = 3 \cdot 5 \cdot 17$	no
9	$511 = 7 \cdot 73$	no
10	$1023 = 3 \cdot 11 \cdot 31$	no

The values of k making M_k prime have been starred in the table. They are $k = 2, 3, 5$, and 7—exactly the primes. The natural conjecture would be that M_k is prime exactly when k is prime. This is soon punctured, however, since

$$M_{11} = 2047 = 23 \cdot 89.$$

(In spite of this, Christian Wolf (1679–1754) stated in print that 2047 was prime. He even claimed that $M_9 = 511$ was prime!)

Half of the conjecture is true. For M_k to be prime, k must be prime. In fact an examination of our table seems to indicate that if $d|k$, then $M_d|M_k$. For example

$$2|4, \quad \text{and} \quad M_2 = 3|15 = M_4.$$

Likewise

$$2|6 \text{ and } 3|6, \quad \text{and} \quad M_2 = 3 \text{ and } M_3 = 7 \text{ both divide } 63 = M_6.$$

Although there are many ways of proving that this works in general (see the problems at the end of the last section), by far the easiest is by means of congruences.

THEOREM   If $d|k$, then $2^d - 1|2^k - 1$. Thus if M_k is prime, k must be prime.

Proof   We wish to show $2^k - 1 \equiv 0 \pmod{2^d - 1}$, or

$$2^k \equiv 1 \pmod{2^d - 1}.$$

Let $k = de$. Then since $2^d \equiv 1 \pmod{2^d - 1}$ we have

$$2^k - 1 = 2^{de} - 1 = (2^d)^e - 1 \equiv 1^e - 1 = 0 \pmod{2^d - 1},$$

where the congruence follows from Theorem 1.12.

The History of Mersenne Primes

The Mersenne numbers are named after Marin Mersenne, a French priest who studied them and published a list (containing a number of mistakes) of primes among them. He also wrote a book on music theory. Over the years more and more Mersenne primes have been found; for example Euler proved that M_{31} was prime. It is an unfortunate characteristic of powers of 2 that they get big quite fast; M_{31} is already more than two billion. The introduction of electronic computers has made it possible to check much larger numbers than before. As late as 1948 the largest k for which M_k was known to be prime was 127, which had been found by the French mathematician Lucas in 1876. It gave the prime

$$M_{127} = 170{,}141{,}183{,}460{,}469{,}231{,}731{,}687{,}303{,}715{,}884{,}105{,}727.$$

By 1953 the American Raphael M. Robinson had used the SWAC computer to find a number of considerably larger primes. His champ was M_{2281}, a number of 687 digits.

These finds were bound to be surpassed as computers got bigger and faster, and they were. The postage-meter stamp shown in Figure 3.1 was used by the University of Illinois to honor the discovery of the prime M_{11213} there in 1963.

Figure 3-1

In 1978 two 18-year-olds from Hayward, California—Laura Nickel and Curt Noll—used 440 hours of computer time to find the 6533-digit prime number M_{21701}. The following story appeared in *The Times* of London.

Prime Number Record Broken

Hayward, California, Nov. 16.—Two 18-year-old American students have discovered with the help of a computer at California State University the biggest known prime number, the number two to the 21,701st power.

Laura Nickel and Curt Noll received congratulations from Dr. Bryant Tuckerman, an American who discovered the previous recordholder among prime numbers: two to the 19,937th power.—Agence France-Presse.

Larger and larger primes continue to be found. The next article appeared in 1985 in the *Washington Post*.

The Washington Post

A 65,000-Digit Number

To check out a new typewriter, one types "The quick brown fox ...." When the computer people at Chevron Geosciences Co. in Houston wanted to check out a new supercomputer intended to analyze geological data, they asked it to run a standard program that looks for prime numbers.

The machine, a $10 million Cray X-MP, obliged. After three hours of performing about 400 million calculations per second, it came up with the largest prime number yet discovered—one that has about 65,000 digits. Expressed in the form that scientists favor, it is 2 raised to the 216,091st power minus 1.

Prime numbers, which are mostly of theoretical interest, are divisible only by themselves and 1. For example, 3 is a prime number but 4, because it can be divided evenly by 2, is not. The ancient Greeks knew there were an infinite number of prime numbers.

The engineers were not expecting to run the program long enough to find a new prime number, but the computer achieved the feat sooner than expected.

"We just happened to crunch enough numbers to come up with a new prime," said William Bartz, a Chevron vice president. "The results are interesting, if true, but they are certainly not going to help me find oil."

—**Boyce Rensberger**

In spite of the many Mersenne primes that have been verified, no one has ever proved that there are infinitely many of them.

Summary of What Is Known and Not Known about Perfect Numbers

1. If $2^k - 1$ is prime, then $2^{k-1}(2^k - 1)$ is a perfect number.
2. If n is an even perfect number, then it is of the form given in (1).
3. It is not known whether there are infinitely many even perfect numbers.
4. It is not known whether there are any odd perfect numbers at all.
5. If $2^k - 1$ is prime, then k is prime, but the converse is not true.

Fermat Numbers

Since the numbers $2^k - 1$ proved interesting, we turn to the numbers $2^k + 1$.

k	$2^k + 1$	prime?
1★	3	yes
2★	5	yes
3	$9 = 3 \cdot 3$	no
4★	17	yes
5	$33 = 3 \cdot 11$	no
6	$65 = 5 \cdot 13$	no
7	$129 = 3 \cdot 43$	no
8★	257	yes
9	$513 = 3 \cdot 3 \cdot 3 \cdot 19$	no
10	$1025 = 5 \cdot 5 \cdot 41$	no

The values of k giving primes have again been starred in the table. So far, they are exactly the powers of 2: 1, 2, 4, and 8. Looking at the composite numbers in the second column we see the same divisors recurring, namely 3 and 5; and these are themselves of the form $2^k + 1$. If k is odd, 3 seems to divide $2^k + 1$, while if k is even (but not divisible by 4), 5 is a divisor.

Our table suggests the conjecture that if $k = ab$, where a is odd, then

$$2^b + 1 \mid 2^k + 1.$$

Although this may be proved by a direct division, a congruence proof is much easier, and provides another demonstration of the usefulness of the congruence notation.

THEOREM If $k = ab$, where a is odd, then

$$2^b + 1 \text{ divides } 2^k + 1.$$

In particular, if $2^k + 1$ is prime, $k > 0$, then k is a power of 2.

Proof We wish to show $2^k + 1 \equiv 0 \pmod{2^b + 1}$, or

$$2^k \equiv -1 \pmod{2^b + 1}.$$

Now $2^b \equiv -1 \pmod{2^b + 1}$, so

$$2^k = 2^{ab} = (2^b)^a \equiv (-1)^a \equiv -1 \pmod{2^b + 1},$$

where the first congruence follows from Theorem 1.12 and the second from the fact that a is odd.

To prove the last sentence of the theorem we note that if k is not a power of 2 then we can take $a > 1$. Thus

$$1 < 2^b + 1 < 2^k + 1,$$

and so $2^k + 1$ has a positive divisor other than 1 and itself.

DEFINITION Fermat numbers
  If r is a nonnegative integer we call

$$2^{2^r} + 1$$

a **Fermat number**, and denote it by F_r.

Examples $F_0 = 2^{2^0} + 1 = 3$, $F_1 = 2^2 + 1 = 5$, $F_2 = 2^4 + 1 = 17$, and $F_3 = 2^8 + 1 = 257$.

Are All Numbers $2^{2^r} + 1$ Prime?

Fermat conjectured that they were. Since 2 raised to a power of 2 is involved, the Fermat numbers get large much faster than even the Mersenne numbers, and so it is much more difficult to tell whether they are prime or not. The number $F_4 = 65537$ is not too big, and can be shown to be a prime, but $F_5 = 2^{32} + 1$ is already greater than 4 billion.

It took about 100 years after Fermat's conjecture for Euler to succeed in showing that F_5 was composite. He used a theoretical method rather than merely checking the 10-digit number for factors.

We will give a congruence proof that 641 divides F_5 involving very little arithmetic. First notice that

$$641 = 640 + 1 = 2^7 5 + 1$$

and

$$641 = 625 + 16 = 5^4 + 2^4.$$

Thus $2^7 5 \equiv -1 \pmod{641}$ and $2^4 \equiv -5^4 \pmod{641}$.

Then (all congruences being modulo 641 and using Theorems 1.8 and 1.12),

$$F_5 = 2^{32} + 1 = 2^{28}2^4 + 1 \equiv (2^7)^4(-5^4) + 1 = -(2^7 5)^4 + 1 \equiv -(-1)^4 + 1 = 0.$$

We see Fermat's conjecture was incorrect. Worse than that, no other primes F_r have as yet been found! Many other values of r are known for which F_r is composite, but aside from these numerical results not much is known. It is possible that F_r is prime only for $r = 0, 1, 2, 3,$ and 4; but it is also possible that other Fermat primes exist, even infinitely many.

Just as Mersenne primes apply to the study of perfect numbers, Fermat primes arise in other parts of mathematics. They are connected with the construction of regular polygons in geometry. For example, the reason that a regular pentagon can be constructed with ruler and compass but a regular 7-sided polygon cannot is that 5 is a Fermat prime but 7 is not. In fact the great German mathematician Gauss proved that a regular polygon with $n > 2$ sides is constructible if and only if n is a power of 2 times a product (possibly empty) of distinct Fermat primes.

PROBLEMS FOR SECTION 3.4

 In the first five problems find the prime power factorization of $2^k + 1$.

    1. $k = 11$ 2. $k = 12$ 3. $k = 13$ 4. $k = 14$ 5. $k = 15$

The next eight problems are based on the material of Sections 3.3 and 3.4. For each statement write T (true), F (false), or N (nobody knows). Give counterexamples for the false statements.

  6. If p is prime, then $2^p - 1$ is prime.
  7. If $2^k(2^{k+1} - 1)$ is perfect, then $2^k - 1$ is prime.
  8. There are only 4 Fermat primes.
  9. There are infinitely many prime Fermat numbers.
 10. Every perfect number is of the form $2^{k-1}(2^k - 1)$.
 11. If $2^k + 1$ is prime, then k is a power of 2.
 12. There are infinitely many odd perfect numbers.
 13. If $d|k$, then $M_d | M_k$.

 Let a and k be positive integers, a odd, and suppose $k = ab$. The next two problems give alternate proofs that $2^b + 1 | 2^k + 1$, by proving the formula $2^k + 1 = (2^b + 1)(2^{b(a-1)} - 2^{b(a-2)} + 2^{b(a-3)} - \cdots + 1).$

 14. Prove the formula by applying the formula for the sum of a geometric progression to the factor on the right.
 15. Prove the formula by multiplying out the right side.

16. Let $a > 0$. Show that if d and e are positive integers, then $a^d - 1$ divides $a^{de} - 1$. (**Hint:** Show $a^{de} \equiv 1 \pmod{a^d - 1}$.)

17. Prove that if $a > 0$, $k > 1$, and $a^k - 1$ is prime, then $a = 2$ and k is prime.

18. Prove that if $a > 1$, $e > 0$, and $q > 0$, q odd, then $a^e + 1$ divides $a^{qe} + 1$. (**Hint:** Show $a^{qe} \equiv -1 \pmod{a^e + 1}$.)

19. Show that if $a > 1$, $k > 0$, and $a^k + 1$ is prime, then a is even and k is a power of 2.

> **C**

20. Show that if $0 < j < k$, then F_j divides $F_k - 2$.

21. Show that if $0 < j < k$, then F_j and F_k are relatively prime.

22. Use the last problem to show there exist infinitely many primes.

23. Prove by induction on n that $F_n = F_0 F_1 \cdots F_{n-1} + 2$ for all positive integers n.

24. Show that no power of a prime is perfect.

25. Show that if $n > 1$, then $\sigma(n)/n < \Pi p/(p - 1)$, where the product is over all primes p dividing n.

26. Show that if n is an odd perfect number, then at least 3 primes divide n.

3.5 The Euler Phi Function

Measuring "Primeness"

There has hardly been anything we have done up to now in which the prime numbers have not played an important role. Although being prime is an all-or-nothing proposition, we have the feeling that some numbers are more composite than others. For example $60 = 2 \cdot 2 \cdot 3 \cdot 5$ seems further from primeness than $62 = 2 \cdot 31$. One way we might quantify this feeling is with $d(n)$, the number of positive divisors of n. Only if n is prime can $d(n)$ be 2; for composite numbers it is bigger. In a certain sense the bigger $d(n)$ is, the further n is from being prime. For example $d(62) = 4$, while $d(60) = 12$.

Another way to measure the same thing depends on the fact that if p is prime, then $(p, n) > 1$ if and only if $p | n$. This property characterizes the primes, for if a is composite it has a divisor n such that $1 < n < a$. Then $(a, n) > 1$ but a doesn't divide n.

Other things being equal, we expect $(a, n) = 1$ to be a more common occurrence when a is prime than when a is composite. As an example let us write out the first few n such that $(7, n) = 1$. The easiest way is to start writing out all the integers, crossing out the multiples of 7.

1	2	3	4	5	6	~~7~~
8	9	10	11	12	13	~~14~~
15	16	17	18	19	20	~~21~~

Let us now do the same thing with 6 instead of 7. Since $(6, n) > 1$ if and only if some prime divides both n and $6 = 2 \cdot 3$, it suffices to cross out the multiples of 2 and 3.

1	2̶	3̶	4̶	5	6̶
7	8̶	9̶	1̶0̶	11	1̶2̶
13	1̶4̶	1̶5̶	1̶6̶	17	1̶8̶

The n such that $(6, n) = 1$ appear to be much rarer.

The reader has probably noticed a pattern in the two arrays just presented. Either all or none of the numbers in each column have been crossed out. This is no surprise if we recall that Theorem 1.11 says that if $b \equiv b'$ (mod a), then $(a, b) = (a, b')$. In particular $(a, b) = 1$ if and only if $(a, b') = 1$.

In the first array we listed the numbers in 7 columns, so those in any given column were all congruent modulo 7. In the second array the columns consisted of numbers congruent modulo 6. By the argument in the previous paragraph, either all the numbers in a given column are relatively prime to a ($=6$ or 7), or else none are.

Since we cannot count all the n such that $(a, n) = 1$, there being infinitely many of them, let us merely count those in the "first row"; after that the pattern repeats anyway.

DEFINITION   Euler φ function

Suppose a is a positive integer. We define $\varphi(a)$ to be the number of integers n, $1 \le n \le a$, such that $(a, n) = 1$. The function φ is called the **Euler phi function**.

Examples   By what we have already seen, $\varphi(6) = 2$ and $\varphi(7) = 6$. Clearly $\varphi(p) = p - 1$ for any prime p. By definition $\varphi(1) = 1$.

Let us compute $\varphi(12)$. Since $12 = 2^2 \cdot 3$, it suffices to cross out the multiples of 2 and 3 from among the first 12 integers:

1	2̶	3̶	4̶	5	6̶	7	8̶	9̶	1̶0̶	11	1̶2̶

We see $\varphi(12) = 4$.

A Shortcut to Computing $\varphi(n)$

Computing $\varphi(12)$ by our "crossing-out" method amounted to eliminating the same numbers as in computing $\varphi(6)$, the multiples of 2 and 3. Thus the numbers left are exactly the same as in the first two rows of our table for 6 above, and the pattern from 7 to 12 repeats that of 1 through 6.

If we had noticed this at the start we could have cut our work in half. We could have crossed out the multiples of 2 and 3 among the first 6 integers, leaving 2 numbers, then doubled this count of 2 to get $\varphi(12) = 4$. In other words, $\varphi(12) = 2\varphi(6)$.

In the same way $\varphi(18) = 3\varphi(6) = 6$. This works since 18 has no prime divisor that does not already divide 6. (Note that $\varphi(30) = 8 \neq 5\varphi(6)$.)

THEOREM A   If all primes dividing k also divide a, then $\varphi(ka) = k\varphi(a)$.

Proof   Consider the array

$$
\begin{array}{cccc}
1 & 2 & \cdots & a \\
a + 1 & a + 2 & \cdots & 2a \\
2a + 1 & 2a + 2 & \cdots & 3a \\
& \vdots & & \\
(k - 1)a + 1 & & \cdots & ka.
\end{array}
$$

Let us cross out the entries that are not relatively prime to ka in order to compute $\varphi(ka)$. This amounts to crossing out everything not relatively prime to a, since $(a, n) > 1$ if and only if some prime divides both a and n, and the same primes divide ka as divide a. By Theorem 1.11 the pattern of crossings-out is the same in each row. Since there are $\varphi(a)$ entries left in the first row, there must be $k \cdot \varphi(a)$ entries left in all. Thus $\varphi(ka) = k \cdot \varphi(a)$.

Example   Since 3 is prime, $\varphi(3) = 3 - 1 = 2$. Thus $\varphi(27) = \varphi(9 \cdot 3) = 9 \cdot \varphi(3) = 9 \cdot 2 = 18$. We can apply Theorem A because 3 is the only prime dividing 9.

A Formula for φ

Theorem A goes a long way toward a formula for $\varphi(n)$. For example, suppose we want to compute $\varphi(1000000)$. We have

$$
\varphi(10^6) = \varphi(2^6 5^6) = \varphi(2^5 5^5 2 \cdot 5) = 2^5 5^5 \cdot \varphi(10),
$$

but we still must compute $\varphi(10)$ by actual count. The computation of $\varphi(n)$ can always be reduced in the same way to that of a product of distinct primes. If we could pull the p_1 out of $\varphi(p_1 p_2 \cdots p_t)$ just as we learned to pull the k out of $\varphi(ka)$ under the hypothesis of Theorem A, we could write a complete formula.

Let us see if we can modify the proof of Theorem A to compute $\varphi(pa)$, where p is a prime *not* dividing a. As before, let us write out the integers from 1 to pa.

$$
\begin{array}{cccc}
1 & 2 & \cdots & a \\
a + 1 & a + 2 & \cdots & 2a \\
& \vdots & & \\
(p - 1)a + 1 & (p - 1)a + 2 & \cdots & pa
\end{array}
$$

Again let us imagine crossing out the n such that $(pa, n) > 1$. The difference here is that $(pa, n) > 1$ is not equivalent to $(a, n) > 1$; the latter implies the former but not conversely. Thus of the $p \cdot \varphi(a)$ integers left after crossing out all n such that $(a, n) > 1$, more still must be eliminated, namely the multiples of p. These are just $1p, 2p, \ldots, ap$. Of these, those kp such that $(k, a) > 1$ have already been eliminated, and there are just $\varphi(a)$ more to cross out in order that $\varphi(pa)$ be properly counted. Thus

$$\varphi(pa) = p \cdot \varphi(a) - \varphi(a) = (p - 1)\varphi(a).$$

We have proved the following theorem:

THEOREM B   If p is prime and p doesn't divide a, then

$$\varphi(pa) = (p - 1)\varphi(a).$$

Examples

$$\varphi(10^6) = 10^5\varphi(2 \cdot 5) = 10^5(2 - 1)\varphi(5) = 400000$$

$$\varphi(60) = \varphi(2^23 \cdot 5) = 2 \cdot \varphi(2 \cdot 3 \cdot 5)$$
$$= 2(2 - 1)\varphi(3 \cdot 5) = 2(3 - 1)\varphi(5) = 16$$

$$\varphi(62) = \varphi(2 \cdot 31) = (2 - 1)\varphi(31) = 30$$

THEOREM 3.4   If $n = p_1^{k_1}p_2^{k_2} \cdots p_t^{k_t}$, where the p's are distinct primes and the k's are positive, then

$$\varphi(n) = \prod_{i=1}^{t} (p_i - 1)p_i^{k_i - 1}.$$

Proof   Let $Q = p_2^{k_2}p_3^{k_3} \cdots p_t^{k_t}$. Then, using Theorems A and B, we have

$$\varphi(n) = \varphi(p_1^{k_1}Q) = p_1^{k_1 - 1}\varphi(pQ) = p_1^{k_1 - 1}(p_1 - 1)\varphi(Q).$$

The same technique may be applied to the prime powers in Q until we arrive at the formula of the theorem.

Examples

$$\varphi(60) = \varphi(2^23 \cdot 5) = (2 - 1)2(3 - 1)(5 - 1) = 16$$

$$\varphi(62) = \varphi(2 \cdot 31) = (2 - 1)(31 - 1) = 30$$

$$\varphi(360) = \varphi(2^33^25) = (2 - 1)2^2(3 - 1)3(5 - 1) = 96$$

The Function Phi Is Multiplicative

It is not hard to prove that φ is multiplicative now that we have a formula for it; in fact we leave this proof for the exercises. Note that this reverses our practice of late of first proving a function multiplicative, and then using this fact to derive a formula.

PROBLEMS FOR SECTION 3.5

For each n given in the first eight problems write out the integers from 1 to n, then cross out those not relatively prime to n. Use this to compute $\varphi(n)$.

1. $n = 4$ 2. $n = 8$ 3. $n = 16$ 4. $n = 12$
5. $n = 3$ 6. $n = 9$ 7. $n = 15$ 8. $n = 10$

9. Which of the following are true by Theorem A?
   (a) $\varphi(3 \cdot 12) = 3\varphi(12)$ (b) $\varphi(4 \cdot 12) = 4\varphi(12)$ (c) $\varphi(8 \cdot 12) = 8\varphi(12)$
   (d) $\varphi(12 \cdot 8) = 12\varphi(8)$ (e) $\varphi(5 \cdot 12) = 5\varphi(12)$ (f) $\varphi(5 \cdot 12) = 4\varphi(12)$
10. Which of the following are true by Theorem B?
    (a) $\varphi(3 \cdot 12) = 3\varphi(12)$ (b) $\varphi(3 \cdot 12) = 2\varphi(12)$
    (c) $\varphi(5 \cdot 12) = 5\varphi(12)$ (d) $\varphi(5 \cdot 12) = 4\varphi(12)$
    (e) $\varphi(25 \cdot 12) = 24\varphi(12)$ (f) $\varphi(35 \cdot 12) = 24\varphi(12)$

In the next eight problems compute $\varphi(n)$ for each n by using the formula.

11. $n = 100$ 12. $n = 200$ 13. $n = 27$ 14. $n = 210$
15. $n = 999$ 16. $n \stackrel{.}{=} 127$ 17. $n = 91$ 18. $n = 1001$

19. This problem illustrates the proof of Theorem B, with $p = 5$ and $a = 6$. Write out $1, 2, \ldots, 30$ in rows of 6.
    (a) Cross out all n such that $(6, n) > 1$. How many integers are left?
    (b) Circle the multiples of 5. How many are there?
    (c) How many multiples of 5 are not already crossed out?
    (d) If the multiples of 5 are crossed out how many of the 30 numbers are not crossed out?
20. Compute $\varphi(n)$ for $n = 100, 200, 27$, and 210, using the formula of problem 22 below.

B 21. Use Theorem 3.4 to prove that φ is multiplicative.
22. Show that for all n

$$\varphi(n) = n \prod_{\substack{p \mid n \\ p \text{ prime}}} \left(1 - \frac{1}{p}\right)$$

In the next four problems find all solutions x to $\varphi(x) = n$ and prove you have found them all. (It has been conjectured that there is no n such that $\varphi(x) = n$ has exactly one solution x, but this has never been proved.)

23. $n = 1$ 　　　　24. $n = 2$ 　　　　25. $n = 4$ 　　　　26. $n = 14$

27. Show that the formula of Theorem 3.4 fails if some of the k's are allowed to be 0.

28. Prove that for all positive integers n

$$\sum_{d|n} \varphi(d) = n.$$

29. Let $\varepsilon > 0$. Prove there exists n such that

$$\left| \frac{\varphi(n)}{n} - 1 \right| < \varepsilon.$$

30. Prove directly from the definition of φ that if p is prime, then $\varphi(p^k) = (p-1)p^{k-1}$.

31. What are all n such that $\varphi(n)$ is odd?

32. What are all n such that $\varphi(2n) = \varphi(n)$?

33. What are all n such that $\varphi(2n) = \varphi(3n)$?

34. Show that if $\varphi(n)$ divides $n - 1$, then the square of no prime divides n.

35. Show that $\varphi(n^2) = n\varphi(n)$ for all positive integers n.

36. Show that if 4 doesn't divide $\varphi(n)$, then n is 1, 2, 4, or of the form p^k, where p is a prime congruent to 3 (mod 4).

37. Show that if n is prime, then $\varphi(n) + \sigma(n) = nd(n)$.

 38. Show that if $\varphi(n) + \sigma(n) = nd(n)$, then n is prime.

39. Show that for all positive integers n

$$\sum_{d|n} M(d)\, \varphi(d) = \prod_{\substack{p|n \\ p \text{ prime}}} (2 - p),$$

where M is as in the problems for Section 3.2.

40. Show that $\varphi'' = \sigma$, in the notation of the problems for Section 3.2.

41. Show that if n is fixed, then $\varphi(x) = n$ has only a finite number of solutions.

42. Find N such that $n \geq N$ implies $\varphi(n) \geq 100$.

43. Show that $\varphi(n) \to \infty$ as $n \to \infty$.

44. Show that if n is such that $\varphi(x) = n$ has exactly one solution x, then 36 divides n.

45. Another proof that there are infinitely many primes uses the function φ. Assume there are only finitely many primes, and let P be their product. Derive a contradiction by computing $\varphi(P)$.

In the next five problems let N(n), E(n), and O(n) be, respectively, the number of solutions, number of even solutions, and number of odd solutions x to $\varphi(x) = 2^n$.

46. Confirm that $N(n) = n + 2$ for $n = 0, 1, 2, 3, 4, 5$. (This formula holds for $n < 32$.)
47. Show that if $n \geq 1$, then $N(n) = E(n - 1) + 2 \cdot O(n)$.
48. Show that $N(n) = n + 2$ for all $n \geq 0$ if and only if $O(n) = 1$ for all $n \geq 0$.
49. Show the odd number x is a solution of $\varphi(x) = 2^n$ if and only if x is the product of distinct Fermat primes.
50. Show that $N(32) = 32$.

3.6 The Möbius Inversion Formula

In Section 3.1 we saw that finding a formula for a numerical function was easier if we knew that it was multiplicative, for in that case we had only to determine its value at prime powers. Theorem 3.3 states that if two numerical functions f and F are related by

$$F(n) = \sum_{d|n} f(d), \quad n = 1, 2, \ldots, \tag{*}$$

then F is multiplicative whenever f is. If f is a simple function, then it may be easy to show it is multiplicative, from which the multiplicativity of F follows. Examples are $f(n) = 1$, $F(n) = d(n)$, and $f(n) = n$, $F(n) = \sigma(n)$.

Given a numerical function F, can we always find a function f so that (*) holds? If so, is f unique? These are questions that will be answered in this section.

The Uniqueness of f

Let us start with as simple an example as we can. Consider the function F defined by

$$F(n) = 1, \quad n = 1, 2, \ldots$$

We will try to find a function f such that (*) holds. Taking $n = 1$ in (*) we get $F(1) = f(1)$, and so $f(1) = F(1) = 1$. Taking $n = 2$ yields $F(2) = f(2) + f(1)$, or $1 = f(2) + 1$, and so we must have $f(2) = 0$. Likewise using $n = 3$ we get $F(3) = f(1) + f(3)$, or $1 = 1 + f(3)$, and so $f(3) = 0$ also. In the same way $F(4) = f(1) + f(2) + f(4)$, or $1 = 1 + 0 + f(4)$, implying that $f(4) = 0$.

It appears that we may have $f(1) = 1$ and $f(n) = 0$ for $n > 1$. Let us not lose sight of our logical position, however. We have *assumed* the existence of a function f satisfying (*) when $F(n)$ is identically 1. Our calculations show that if such a function f exists, then of necessity $f(1) = 1$, and $f(2) = f(3) = f(4) = 0$. Such computations do not establish the existence of such a function, however.

An analogous situation is the way many people would "solve" the equation

$$3x + 2 = -4,$$

by writing

$$3x = -4 - 2 = -6,$$

$$x = -\frac{6}{3} = -2.$$

This sequence of computations shows that if $3x + 2 = -4$ has a solution, then it must be $x = -2$. After all, what is "x" that is manipulated in the equations? It must be a number assumed to satisfy the original equation. We must substitute -2 for x in the original equation to actually show that -2 is a solution.

The problem at hand may be handled in the same way. Let us define the function f by

$$f(n) = \begin{cases} 1, & n = 1 \\ 0, & n > 1. \end{cases}$$

Then it is almost obvious that

$$\sum_{d \mid n} f(d) = 1 \quad \text{for all positive integers } n,$$

and so f is a solution to our problem.

Of course our calculation of $f(n)$ for $n = 1, 2, 3,$ and 4 was not worthless; it suggested to us what function to try for f. It also indicated that if an f satisfying (∗) exists, then its values are determined by those of F, and so it is unique. A formal proof of this fact requires the use of Induction II.

THEOREM Suppose F is a numerical function. Then there is at most one function f such that

$$F(n) = \sum_{d \mid n} f(d) \tag{∗}$$

for all positive integers n.

Proof *(Induction II)* Suppose that g is a function such that also

$$F(n) = \sum_{d \mid n} g(d) \tag{∗∗}$$

for all positive integers n. We will show by induction on n that $f(n) = g(n)$, $n = 1, 2, \ldots$.

1. Taking $n = 1$ in $(*)$ and $(**)$ we see that $F(1) = f(1)$ and $F(1) = g(1)$, so $f(n) = g(n)$ for $n = 1$.
2. Suppose for some integer k we have $f(n) = g(n)$ for $n = 1, 2, \ldots, k$. Then

$$F(k + 1) = \sum_{d|k+1} f(d) = \sum_{d|k+1} g(d).$$

By the induction hypothesis the two sums on the right have all their terms equal except possibly $f(k + 1)$ and $g(k + 1)$, which must therefore also be equal.

By the Induction II principle we see that $f(n) = g(n)$ for all positive integers n.

The Möbius Function

In the above example (with F identically 1) we computed a few values that a function f had to take on in order to satisfy $(*)$, then took a stab at what f was in general. Fortunately the function f we were after was sufficiently simple for a natural guess to be made, but this is too much to hope for in general. Had we known that f would be multiplicative, then determining what it had to be would have been simpler. Recall we knew $f(1)$ was forced to be 1. If p is any prime number, then $(*)$ implies that $F(p) = f(1) + f(p)$, or $1 = 1 + f(p)$, and so $f(p) = 0$. Likewise $F(p^2) = f(1) + f(p) + f(p^2)$, or $1 = 1 + 0 + f(p^2)$, and so $f(p^2) = 0$ also. Continuing in this way we see that $f(p^k) = 0$ for all positive integers k. Then if $n > 1$ we have $n = p^k m$ for some prime p not dividing m. Thus if f is multiplicative we have $f(n) = f(p^k)f(m) = 0$.

It turns out that an Induction II proof similar to that of the last theorem can be used to prove that if numerical functions f and F satisfy $(*)$, and if F is multiplicative (as it was in our example), then f must be multiplicative also. This is the converse of Theorem 3.3.

THEOREM   Suppose f and F are numerical functions satisfying

$$F(n) = \sum_{d|n} f(d), \quad n = 1, 2, 3, \ldots \tag{$*$}$$

Then if F is multiplicative, so is f.

Proof *(Induction II)*   Let a and b be relatively prime positive integers. We must show that $f(ab) = f(a)f(b)$. The proof will be by induction on ab.

1. If $ab = 1$, then $a = b = 1$. From $(*)$ we have $F(1) = f(1)$. But then $F(ab) = F(a)F(b)$ implies the same equation for f, since $ab = a = b = 1$.

2. Suppose we know that $f(ab) = f(a)f(b)$ whenever a and b are relatively prime and $ab = 1, 2, 3, \ldots, k$. Consider $ab = k + 1$. We have

$$F(ab) = \sum_{c|ab} f(c) =_{\mathbf{1}} \sum_{\substack{d|a \\ e|b}} f(de) = f(ab) + \sum_{\substack{d|a \\ e|b}}{}' f(de) =_{\mathbf{2}} f(ab) + \sum_{\substack{d|a \\ e|b}}{}' f(d)f(e),$$

where the prime on the last two summations indicates that the term with $d = a$ and $e = b$ has been omitted from each. The equation marked **1** is justified by Theorem 2.5, and the equation marked **2** follows from the induction hypothesis and the fact that if $(a, b) = 1$ and $d|a$ and $e|b$, then $(d, e) = 1$ also.

On the other hand we have

$$F(a)F(b) = \sum_{d|a} f(d) \sum_{e|b} f(e) = \sum_{\substack{d|a \\ e|b}} f(d)f(e) = f(a)f(b) + \sum_{\substack{d|a \\ e|b}}{}' f(d)f(e),$$

where the term $f(a)f(b)$ has been left out of the last summation. But $F(ab) = F(a)F(b)$ and so we see $f(ab) = f(a)f(b)$ when $ab = k + 1$.

Thus by Induction II $f(ab) = f(a)f(b)$ for all pairs of relatively prime positive integers a and b.

Example   Suppose $F(1) = 1$, while if $n > 1$ then $F(n) = 0$. (Thus we are taking F to be the function called f in the previous example.) Suppose there exists a function f satisfying (∗). Find a formula for f.

It is easy to show directly that the function F is multiplicative. (This could also be derived from the last theorem and the fact that the function that is identically 1 is multiplicative.) Thus by the last theorem the function f we are after is also multiplicative. We will use this fact to determine f.

As usual $f(1) = F(1) = 1$. Let p be prime. Then from (∗) $F(p) = f(1) + f(p)$, or $0 = 1 + f(p)$, from which we see that $f(p) = -1$. Likewise $F(p^2) = f(1) + f(p) + f(p^2)$, or $0 = 1 - 1 + f(p^2)$, and so $f(p^2) = 0$. In the same way $F(p^3) = f(1) + f(p) + f(p^2) + f(p^3)$, or $0 = 1 - 1 + 0 + f(p^3)$, so $f(p^3) = 0$ also. Continuing in this way we see that $f(p^k) = 0$ for all integers $k > 1$.

Now we use the multiplicativity of f. Let $n = p_1^{r_1} p_2^{r_2} \cdots p_t^{r_t}$, the p's distinct primes and the r's positive integers. Then $f(n) = f(p_1^{r_1}) \cdots f(p_t^{r_t})$. If any exponent r_i is greater than 1 then $f(p_i^{r_i}) = 0$, and so $f(n) = 0$ also. On the other hand if $r_1 = r_2 = \cdots = 1$, then $f(n) = (-1)(-1) \cdots (-1) = (-1)^t$. The function f we have found is important in number theory and is conventionally denoted by μ.

DEFINITION   Möbius function

The **Möbius function** is the function μ defined by

$$\mu(n) = \begin{cases} 1 & \text{if } n = 1, \\ 0 & \text{if } p^2|n \text{ for some prime } p, \\ (-1)^t & \text{if } n = p_1 p_2 \cdots p_t, \text{ the } p\text{'s distinct primes.} \end{cases}$$

For example $\mu(28) = 0$ since $2^2 | 28$, $\mu(29) = (-1)^1 = -1$, and $\mu(30) = (-1)^3 = -1$.

Since in our last example it was assumed that a function f existed satisfying (*) for $F(1) = 1$ and $F(n) = 0$ for $n > 1$, we must check that μ really works.

THEOREM   The function μ is multiplicative, and

$$\sum_{d|n} \mu(d) \tag{***}$$

is 1 if $n = 1$ and 0 if $n > 1$.

Proof   First we show μ is multiplicative. Suppose $(a, b) = 1$. We wish to show that $\mu(ab) = \mu(a)\mu(b)$. If either a or b is 1 this is easy. Likewise if the square of some prime divides a or b, then both sides are 0. The only remaining case is when $a = p_1 \cdots p_t$ and $b = q_1 \cdots q_u$, the p's and q's distinct primes. Then no p is a q since $(a, b) = 1$. Thus $\mu(ab) = \mu(p_1 \cdots p_t q_1 \cdots q_u) = (-1)^{t+u} = (-1)^t(-1)^u = \mu(a)\mu(b)$.

Now let $G(n)$ stand for the summation of (***). Since we have proved that μ is multiplicative, so is G by Theorem 3.3. But by (***) $G(1) = \mu(1) = 1$, while if p is prime,

$$G(p^k) = \mu(1) + \mu(p) + \mu(p^2) + \cdots + \mu(p^k) = 1 - 1 + 0 + \cdots + 0 = 0$$

for $k > 0$, and so since G is multiplicative $G(n) = 0$ for $n > 1$.

The Möbius Inversion Formula

Now we consider the general problem of finding a function f satisfying (*) for all positive integers n, given a numerical function F. Let us try to work out how the values of f must depend on those of F. Taking $n = 1$ in (*) tells us that we must have $f(1) = F(1)$. Now let p be prime. Then by (*) with $n = p$ we have $F(p) = f(1) + f(p)$, or $F(p) = F(1) + f(p)$, and so

$$f(p) = F(p) - F(1). \tag{1}$$

Likewise taking $n = p^2$ yields

$$F(p^2) = f(1) + f(p) + f(p^2) = F(1) + (F(p) - F(1)) + f(p^2),$$

and so

$$f(p^2) = F(p^2) - F(p). \tag{2}$$

The reader should make a similar computation to check that we must have

$$f(p^3) = F(p^3) - F(p^2). \tag{3}$$

Note that equations (1), (2), and (3) can all be expressed as

$$f(n) = F(n) - F\left(\frac{n}{p}\right) \text{ for } n = p, p^2, \text{ or } p^3, \tag{4}$$

and we might suspect that this formula holds for n any power of a prime.

Now let us see what happens when more than one prime is involved, say $n = pq$, where p and q are distinct primes. From (∗) and (1) we see

$$F(pq) = f(1) + f(p) + f(q) + f(pq)$$
$$= F(1) + (F(p) - F(1)) + (F(q) - F(1)) + f(pq),$$

and so $f(pq) = F(pq) - F(p) - F(q) + F(1)$, which we could write

$$f(n) = F(n) - F\left(\frac{n}{p}\right) - F\left(\frac{n}{q}\right) + F\left(\frac{n}{pq}\right) \text{ for } n = pq. \tag{5}$$

The reader should check that (5) also holds for $n = p^2q$, and also that if p, q, and r are distinct primes and $n = pqr$, then

$$f(n) = F(n) - F\left(\frac{n}{p}\right) - F\left(\frac{n}{q}\right) - F\left(\frac{n}{r}\right) + F\left(\frac{n}{pq}\right) \tag{6}$$
$$+ F\left(\frac{n}{pr}\right) + F\left(\frac{n}{qr}\right) - F\left(\frac{n}{pqr}\right).$$

It appears we can express $f(n)$ in the form $\sum m(d)F(n/d)$, where d runs through the positive divisors of n, and m is a function that takes on only the values 0, 1, and -1. In fact it appears that $m(d) = 0$ unless d is 1 or a product of distinct primes, and in the latter case $m(d)$ is -1 or 1 based on whether an odd or even number of primes is involved. But this exactly describes the Möbius function.

THE MÖBIUS INVERSION FORMULA   Let F be any numerical function. Define f by

$$f(n) = \sum_{d|n} \mu(d)F\left(\frac{n}{d}\right), \quad n = 1, 2, 3, \ldots.$$

Then

$$F(n) = \sum_{d|n} f(d) \quad \text{for all integers } n.$$

Proof   To prove this theorem we must substitute the function f defined by the first equation into the right side of the second one and confirm that we actually get $F(n)$ for all positive integers n. This substitution gives

$$\sum_{d|n}\left(\sum_{e|d} \mu(e)F\left(\frac{d}{e}\right)\right).$$

Notice that since $d|n$ the number d/e is always some divisor k of n. We will regroup this sum as

$$\sum_{k|n} h(k)F(k)$$

by collecting the terms containing $F(k)$ for a fixed divisor k of n. For example if $n = 4$ we could write

$$\sum_{d|4}\left(\sum_{e|d}\mu(e)F\left(\frac{d}{e}\right)\right) = \sum_{e|1}\mu(e)F\left(\frac{1}{e}\right) + \sum_{e|2}\mu(e)F\left(\frac{2}{e}\right) + \sum_{e|4}\mu(e)F\left(\frac{4}{e}\right)$$

$$= \mu(1)F(1) + \mu(1)F(2) + \mu(2)F(1) + \mu(1)F(4) + \mu(2)F(2) + \mu(4)F(1)$$

$$= (\mu(1) + \mu(2) + \mu(4))F(1) + (\mu(1) + \mu(2))F(2) + \mu(1)F(4)$$

$$= (1 - 1 + 0)F(1) + (1 - 1)F(2) + 1F(4) = F(4).$$

In general the coefficient $h(k)$ of $F(k)$ will be $\sum \mu(e)$, where the sum runs over all pairs d, e such that $d|n$, $e|d$, and $d/e = k$. Of course if d/e is the integer k, then it is automatic that $e|d$, and so we can forget this condition. Also once we choose e the equation $d/e = k$ completely determines d; we need only figure out which values of e make $d = ek$ a divisor of n. But $ek|n$ if and only if n/ek is an integer, or e is a divisor of n/k. Thus

$$\sum_{d|n}\left(\sum_{e|d}\mu(d)F\left(\frac{n}{d}\right)\right) = \sum_{k|n}h(k)F(k) = \sum_{k|n}\left(\sum_{e|n/k}\mu(e)\right)F(k).$$

But by the previous theorem the inner sum is 1 when $n/k = 1$ and so $k = n$, and 0 otherwise. Thus this sum is simply $F(n)$, which is what we wanted to prove.

Note that this theorem does not depend on whether the function F is multiplicative or not.

Example   Find a formula for the numerical function f such that

$$\sum_{d|n}f(d) = n$$

for all positive integers n.

Since the function defined by $F(n) = n$ for all $n > 0$ is multiplicative, f will be multiplicative also. Knowing this makes finding f easier. By the Möbius Inversion Formula we have

$$f(n) = \sum_{d|n}\mu(d)\left(\frac{n}{d}\right), \quad n = 1, 2, \dots.$$

Thus $f(1) = \mu(1) = 1$ and if p is prime

$$f(p^k) = \mu(1)p^k + \mu(p)p^{k-1} + \mu(p^2)p^{k-2} + \cdots + \mu(p^k)(1)$$

$$= p^k - p^{k-1} = (p - 1)p^{k-1}$$

for any positive integer k.

Now by the fact that f is multiplicative if $n = p_1^{k_1} p_2^{k_2} \cdots p_t^{k_t}$, the p's distinct primes, we have

$$f(n) = (p_1 - 1)p_1^{k_1 - 1} \cdots (p_t - 1)p_t^{k_t - 1}.$$

But this is exactly the formula for $\varphi(n)$ given by Theorem 3.4.

THEOREM 3.5　For any positive integer n we have

$$\sum_{d|n} \varphi(d) = n.$$

PROBLEMS FOR SECTION 3.6

1. Compute $\mu(n)$ for $n = 1, 2, \ldots, 12$.
2. Compute $\mu(n)$ for $n = 13, 14, \ldots, 24$.
3. Find all n, $25 < n < 50$, such that $\mu(n) = 1$.
4. Find all n, $50 < n < 75$, such that $\mu(n) = 0$.
5. Find all nonprimes $n < 100$ with $\mu(n) = -1$.
6. Find all $n < 100$ such that $\mu(n)\mu(n + 1)\mu(n + 2) = 1$.
7. Confirm that the summation (∗∗∗) of this section is 0 for $n = 12$.
8. Confirm that the summation (∗∗∗) of this section is 0 for $n = 30$.
9. Confirm Theorem 3.5 for $n = 12$.
10. Confirm Theorem 3.5 for $n = 30$.

In the next four problems a function F is given. Find a formula for the function f satisfying equation (∗) of this section.

11. $F(n) = \mu(n)$.
12. $F(n) = (\mu(n))^2$.
13. $F(n) = 1$ if n is odd and 0 if n is even.
14. $F(n) = n^2$.

The next two problems should be done without assuming any results of this section.

15. Suppose $F(n) = 1$ for all positive integers n, and that f is a function satisfying (∗) of this section. Prove by induction on k that if p is prime then $f(p^k) = 0$ for all positive integers k.
16. Suppose $F(1) = 1$ and $F(n) = 0$ for $n > 1$, and that f is a function satisfying (∗) of this section. Prove by induction on k that if p is prime then $f(p^k) = 0$ for all integers $k > 1$.

17. Show that if $n = p_1^{k_1} \cdots p_t^{k_t}$, the p's distinct primes and the k's positive integers, then the formula (∗∗∗) of this section has exactly 2^t nonzero terms.

18. Show that if F and f are functions satisfying (∗), then

$$f(n) = \sum_{d|n} \mu\left(\frac{n}{d}\right) F(d) \quad \text{for all positive integers } n.$$

19. Show that if n is any positive integer, then $\mu(n)\mu(n+1)\mu(n+2)\mu(n+3) = 0$.
20. Prove that there are infinitely many integers n such that $\mu(n) + \mu(n+1) = 0$.
21. Prove that there are infinitely many integers n such that $\mu(n) + \mu(n+1) + \mu(n+2) = 0$.
22. If $n = p_1^{k_1} \cdots p_t^{k_t}$, the p's distinct primes and the k's positive integers, define $F(n)$ to be $(-1)^t$. Find the function f satisfying (∗) of this section.
23. Define $F(n)$ to be 1 if $n = m^2$ for some integer m and 0 otherwise. Find the function f satisfying (∗) of this section.
24. Prove that there exist infinitely many n such that $\mu(n) + \mu(n+1) = -1$.

*In the following problems $[X]$ denotes the greatest integer $\leq X$, called the **integer part** of X, where X is any real number.*

25. Show that if X is a positive real number and p and q are distinct primes, then the number of positive integers $\leq X$ not divisible by p is $[X] - [X/p]$, and the number not divisible by p or q is $[X] - [X/p] - [X/q] + [X/pq]$.
26. Let $p_1, p_2, \ldots, p_t$ be distinct primes and X a positive real number. Show that the number of positive integers $\leq X$ not divisible by any of $p_1, \ldots, p_t$ is

$$[X] - \left[\frac{X}{p_1}\right] - \left[\frac{X}{p_2}\right] - \cdots - \left[\frac{X}{p_t}\right] + \left[\frac{X}{p_1 p_2}\right] + \left[\frac{X}{p_1 p_3}\right] + \cdots$$

$$+ \left[\frac{X}{p_{t-1} p_t}\right] - \left[\frac{X}{p_1 p_2 p_3}\right] - \cdots + (-1)^t \left[\frac{X}{p_1 \cdots p_t}\right].$$

27. Let $n = p_1^{k_1} \cdots p_t^{k_t}$, where the p's are distinct primes and the k's positive integers. Show that the number of positive integers $\leq X$ relatively prime to n is $\sum \mu(d)[X/d]$, where the summation is over the positive divisors d of n.
28. Show that the sum in the previous problem has at most 2^t nonzero terms.
29. Let $\pi(X)$ denote the number of primes $\leq X$. Show that if p_i is the ith prime (so $p_1 = 2$, $p_2 = 3$, etc.), if $p_t < X$, and if $n = p_1 p_2 \cdots p_t$, then $\pi(X) \leq t + \sum \mu(d)[X/d]$, where in the summation d runs through the positive divisors of n.
30. Show that with the hypotheses of the previous problem

$$\pi(X) \leq t + 2^t + X\left(1 - \frac{1}{p_1}\right)\left(1 - \frac{1}{p_2}\right) \cdots \left(1 - \frac{1}{p_t}\right).$$

31. Show that $\pi(X)/X \to 0$ as $X \to \infty$. (**Hint:** See the last problem in Section 2.5.)
32. Show that if f is multiplicative and m and n are positive integers, then $f(m)f(n) = f((m, n))f([m, n])$.

Leonhard Euler
1703–1783

"Prolific" is the word most often applied to the Swiss mathematician Euler, from whom gushed forth a steady flow of work from the age of 19 on, even though he was blind for the last 17 years of his life. (He also had 13 children.) Mainly known for his work in analysis, Euler wrote a calculus textbook and

introduced the present-day symbols for e, π, and i. (He also discovered the wonderful equation linking them: $e^{\pi i} + 1 = 0$.)

Among Euler's discoveries in number theory is the law of quadratic reciprocity, which connects the solvability of the congruences

$$x^2 \equiv p \ (\text{mod } q) \qquad \text{and} \qquad y^2 \equiv q \ (\text{mod } p),$$

where p and q are distinct primes, although it remained for Gauss to provide the first proof.

Most of Euler's life was spent in the courts of Frederick the Great of Prussia and Catherine the Great of Russia,

LEONHARD EULER

in the company of such luminaries as Voltaire. His phenomenal memory (he knew Virgil's *Aeneid* by heart) helped him continue his mathematical researches after he became blind, when he dictated his papers to his sons.

Chapter 4

The Algebra of Congruence Classes

4.1 Solving Linear Congruences

What Does It Mean To Solve a Congruence?

The reader knows what it means to solve an equation. To "solve" the equation

$$x^2 + x - 2 = 0,$$

for example, we must find all values of x making the equation true. We will consider the similar problem where the equality sign is replaced by "$\equiv$." In general we will restrict ourselves to congruences of the form

$$F(x) \equiv 0 \pmod{b},$$

where $F(x)$ is a polynomial in x with integer coefficients.

Suppose x_1 satisfies the congruence

$$x^2 + x - 2 \equiv 0 \pmod{b},$$

and suppose that $x_2 \equiv x_1 \pmod{b}$. Then using Theorem 1.12 and various parts of Theorem 1.8 we have

$$x_2^2 \equiv x_1^2 \pmod{b},$$

$$x_2^2 + x_2 \equiv x_1^2 + x_1 \pmod{b}, \quad \text{and}$$

$$x_2^2 + x_2 - 2 \equiv x_1^2 + x_1 - 2 \pmod{b};$$

and so x_2 is also a solution to the congruence. Thus if we find one solution x we immediately have infinitely many more solutions, namely, all integers congruent to x modulo b. Since we cannot write all of these down we will content ourselves with identifying any one of them, and the one we identify will serve as a representative for the whole class of solutions congruent modulo b.

Clearly the same argument works for any polynomial with integral coefficients.

THEOREM   If F is a polynomial with integral coefficients and if $x_2 \equiv x_1$ (mod b), then $F(x_2) \equiv F(x_1)$ (mod b). In particular, x_2 is a solution to $F(x) \equiv 0$ (mod b) if and only if x_1 is.

It is convenient at this point to introduce some language relating to various sets of integers of importance for a given modulus.

DEFINITIONS   congruence class, complete residue system

Suppose $b > 0$. By a **congruence class modulo** b we mean the set of all integers congruent modulo b to some fixed integer k. By a **complete residue system modulo** b we mean a set of integers such that exactly one integer in the set is in each congruence class modulo b.

Examples   There are 3 congruence classes modulo $b = 3$, namely

$$\{\ldots, -6, -3, 0, 3, 6, 9, \ldots\},$$

$$\{\ldots, -5, -2, 1, 4, 7, 10, \ldots\}, \quad \text{and}$$

$$\{\ldots, -4, -1, 2, 5, 8, 11, \ldots\}.$$

One complete residue system modulo 3 is $\{0, 1, 2\}$. Another is $\{1, 2, 3\}$. Another is $\{10, -1, 6\}$.

If we write out the integers in a table with rows of length b, as has proved useful several times, then the columns are exactly the congruence classes modulo b. Taking $b = 4$, for example, we get

	...		
...	-6	-5	-4
-3	-2	-1	0
1	2	3	4
5	6	7	8
9	10	11	...
	...		,

and each column is a congruence class modulo 4. If we choose exactly one number from each column, say 5, -2, 11, and 0, we get a complete residue system modulo 4.

It is easy to see that there are exactly b congruence classes modulo b, and that any complete residue system modulo b also consists of exactly b elements.

Counting Solutions to a Congruence

Recall that if F is a polynomial with integral coefficients, then any integer congruent modulo b to a solution of

$$F(x) \equiv 0 \pmod{b} \qquad (*)$$

is also automatically a solution, and so doesn't deserve separate attention. This is the reason for the following definition.

DEFINITIONS complete solution, least complete solution

We call $x_1, x_2, \ldots, x_k$ a **complete solution** to the congruence $(*)$ in case it is the set of all solutions in some complete residue system modulo b. It is the **least complete solution** in case it is the set of all solutions among $0, 1, \ldots, b - 1$.

Examples Consider the congruence

$$4x - 18 \equiv 0 \pmod{6}.$$

Among the integers 1, 2, 3, 4, 5, 6 exactly 3 and 6 satisfy the congruence. Thus $x = 3, 6$ is a complete solution. Another complete solution is $x = -3, 0$, since these are all solutions in the complete residue system $-3, -2, -1, 0, 1, 2$. Another is $x = 9, 72$. The least complete solution is $x = 0, 3$.

Note that it is possible to match up the elements of any two complete solutions so that corresponding numbers are congruent modulo b. For example

$$9 \equiv 3 \quad \text{and} \quad 72 \equiv 6 \pmod{6}.$$

On the other hand $x = -3, 9$ is not a complete solution to $4x - 18 \equiv 0 \pmod{6}$ since -3 and 9 are in the same congruence class.

The Congruence $ax \equiv c$ (mod b)

In this chapter we will consider only congruences $F(x) \equiv 0 \pmod{b}$ where F is a linear polynomial. Such a congruence can be put in the form $ax \equiv c \pmod{b}$. To solve this congruence we must look for values of x such that

$$b \mid c - ax,$$

that is, such that

$$c - ax = by$$

for some integer y. This is the equation $ax + by = c$ that we analyzed completely in Section 1.4. There we decided that

1. The equation $ax + by = c$ has a solution if and only if (a, b) divides c.

2. A solution can be found explicitly by using the Euclidean Algorithm to find (a, b) and then solving the equations backwards.
3. If x_0, y_0 is a particular solution, then all solutions are given by $x = x_0 + bt/d$, $y = y_0 - at/d$, where $d = (a, b)$ and t runs through the integers.

We will illustrate the use of (1) and (2) to find a solution to

$$147x \equiv 77 \quad (\text{mod } 161).$$

Here $a = 147$, $b = 161$, and $c = 77$. First we use the Euclidean Algorithm to find $(147, 161)$.

$$161 = 1 \cdot 147 + 14$$

$$147 = 10 \cdot 14 + 7$$

$$14 = 2 \cdot 7 + 0$$

We see that $(147, 161) = 7$, and since 7 divides 77, the congruence has a solution. Now we solve the equations backwards as follows:

$$7 = 147 - 10 \cdot 14$$
$$= 147 - 10(161 - 147)$$
$$= 11 \cdot 147 - 10 \cdot 161.$$

Multiplying through by $77/7 = 11$ gives

$$77 = 121 \cdot 147 - 110 \cdot 161.$$

We have found one solution to the congruence, namely $x = 121$.

Now let us see how (3) tells us how to find a complete solution to the congruence

$$ax \equiv c \quad (\text{mod } b).$$

We know that if x_0 is one solution to the corresponding equation, then all solutions are given by

$$x = x_0 + \frac{bt}{d},$$

where $d = (a, b)$ and t runs through the integers. This gives infinitely many values of x, but we only want to find the solutions in some complete residue system modulo b; that is, we want at most one solution in any congruence class modulo b.

How is it possible for us to have

$$x_0 + \frac{bt}{d} \equiv x_0 + \frac{bt'}{d} \pmod{b}$$

for integers t and t'? This amounts to the fact that b divides

$$x_0 + \frac{bt'}{d} - \left(x_0 + \frac{bt}{d}\right) = \frac{(t'-t)b}{d}.$$

Then $(t'-t)b/d = kb$, or $(t'-t) = kd$, for some integer k; that is,

$$t' \equiv t \pmod{d}.$$

Thus using only values of t incongruent modulo d will certainly give us solutions x incongruent modulo b. Furthermore, we will not miss any solutions this way, for it can be shown that

$$t' \equiv t \pmod{d}$$

implies that

$$x_0 + \frac{bt}{d} \equiv x_0 + \frac{bt'}{d} \pmod{b};$$

see the exercises at the end of this section.

We see that if the values of t are restricted to lie in some complete residue system modulo $d = (a, b)$, then the values of $x = x_0 + bt/d$ will be incongruent modulo b, and so will comprise a complete solution.

THEOREM 4.1   Consider the congruence

$$ax \equiv c \pmod{b}.$$

This has a solution if and only if (a, b) divides c. If x_0 is any solution then a complete solution is given by the numbers

$$x = x_0 + \frac{bt}{(a, b)},$$

where t runs through any complete residue system modulo (a, b).

Example   We return to the congruence

$$147x \equiv 77 \pmod{161},$$

for which we have already found a particular solution $x_0 = 121$. We also found that $(a, b) = (147, 161) = 7$. Then a complete solution to the congruence is given by

$$x = 121 + \frac{161t}{7} = 121 + 23t,$$

where t runs through any complete residue system modulo 7. Letting $t = 0, 1, 2, 3, 4, 5, 6$ yields the complete solution

$$x = 121, 144, 167, 190, 213, 236, 259.$$

(Note that taking $t = 7$ gives $x = 282$; but $282 \equiv 121 \pmod{161}$.) The least complete solution is $x = 121, 144, 167 - 161 = 6, 190 - 161 = 29, 213 - 161 = 52, 236 - 161 = 75, 259 - 161 = 98$.

Example Suppose we are asked for a complete solution to

$$15x \equiv 50 \pmod{100}.$$

By inspection we find the particular solution $x_0 = 10$. We also note that $(15, 100) = 5$. Then a complete solution is given by

$$x = 10 + \frac{100t}{5} = 10 + 20t,$$

where $t = 0, 1, 2, 3, 4$. We get

$$x = 10, 30, 50, 70, 90$$

as a complete solution. (It is also the least complete solution.)

Example Find a complete solution to the congruence

$$1485x \equiv 999 \pmod{2222}.$$

Here

$$2222 = 1 \cdot 1485 + 737$$

$$1485 = 2 \cdot 737 + 11$$

$$737 = 67 \cdot 11 + 0,$$

so $(1485, 2222) = 11$. But 11 doesn't divide 999 and so there are no solutions.

Example Find a complete solution to

$$45x \equiv 3 \quad (\text{mod } 48),$$

where the solutions are to be positive but as small as possible.

We note by inspection the solution $x_0 = -1$. Since $(45, 48) = 3$, a complete solution is

$$x = -1 + \frac{48t}{3} = -1 + 16t,$$

where $t = 0, 1, 2$. This gives

$$x = -1, 15, 31.$$

Now 15 and 31 are the smallest positive elements of their congruence classes modulo 48, but to fulfill the conditions of the problem we must replace -1 by $-1 + 48 = 47$. Thus the answer is

$$x = 15, 31, 47.$$

PROBLEMS FOR SECTION 4.1

In the first eight problems tell whether each sequence given is a complete residue system modulo 6 or not.

1. $-3, -2, -1, 1, 2, 3$
2. $0, 3, 6, 9, 12, 15$
3. $0, 5, 10, 15, 20, 25$
4. $0, 1, 2, 3, 4, 5, 6$
5. $1, 2, 4, 8, 3, 5$
6. $3, 8, 7, 18, 23, 22$
7. $7, 14, 21, 28, 35$
8. $5, -5, 4, -4, 3, 60$

In the next six problems tell whether each sequence forms a complete solution to the given congruence.

9. $x = 4; 2x \equiv 8 \quad (\text{mod } 6)$
10. $x = 4, 10; 2x \equiv 8 \quad (\text{mod } 6)$
11. $x = 4; 2x \equiv 8 \quad (\text{mod } 9)$
12. $x = 4, 10; 10x \equiv 4 \quad (\text{mod } 12)$
13. $x = 0; 6x \equiv 4 \quad (\text{mod } 12)$
14. $x = 3, 7; 3x \equiv 9 \quad (\text{mod } 12)$

In the next six problems tell the number of elements in a complete solution.

15. $5x \equiv 100 \quad (\text{mod } 55)$
16. $11x \equiv 14 \quad (\text{mod } 23)$
17. $91x \equiv 169 \quad (\text{mod } 143)$
18. $91x \equiv 169 \quad (\text{mod } 140)$
19. $1001x \equiv 143 \quad (\text{mod } 99)$
20. $48x \equiv 128 \quad (\text{mod } 1000)$

In the next ten problems give the least complete solution to the congruence.

21. $6x \equiv 2 \pmod 8$ 22. $36x \equiv 30 \pmod{42}$
23. $25x \equiv 100 \pmod{35}$ 24. $143x \equiv 169 \pmod{110}$
25. $27x \equiv -18 \pmod{15}$ 26. $51x \equiv 0 \pmod{17}$
27. $3x \equiv 18 \pmod{18}$ 28. $253x \equiv 341 \pmod{299}$
29. $165x \equiv 84 \pmod{221}$ 30. $441x \equiv 465 \pmod{640}$

31. Find all x, $0 \le x \le 9$, such that $5x \equiv 15 \pmod{10}$.
32. Find all x, $|x| < 5$, such that $5x \equiv 20 \pmod{10}$.
33. Find all x, $100 \le x < 110$, such that $6x \equiv 2 \pmod{10}$.
34. Find all x, $100 \le x < 110$, such that $6x \equiv 9 \pmod{15}$.

B   *In the next three problems you may want to invoke the **pigeon-hole principle**: if $n + 1$ objects are put into n boxes, then some box must contain more than one object. Assume $b > 0$.*

35. Show that if S and T are two congruence classes modulo b, then either $S = T$ or else S and T have no elements in common.
36. Show that if S has b elements and if no two elements of S are congruent modulo b, then S is a complete residue system modulo b.
37. Show that if S has b elements and if S contains an element in each congruence class modulo b, then S is a complete residue system modulo b.

True-False. In the next seven problems state whether each statement is true or false. Give counterexamples for the false statements. Assume that a, b, and c are positive, $(a, b) = 1$, and that $x_1, x_2, \ldots, x_b$ is a complete residue system modulo b.

38. $x_1 + c, x_2 + c, \ldots, x_b + c$ is a complete residue system modulo b.
39. $cx_1, cx_2, \ldots, cx_b$ is a complete residue system modulo b.
40. $ax_1, ax_2, \ldots, ax_b$ is a complete residue system modulo b.
41. $-x_1, -x_2, \ldots, -x_b$ is a complete residue system modulo b.
42. There exists a complete residue system modulo b, say $y_1, y_2, \ldots, y_b$, such that $|y_i| \le b/2$, $i = 1, 2, \ldots, b$.
43. There exists a complete residue system modulo b, say $y_1, y_2, \ldots, y_b$, such that $|y_i| < b/2$, $i = 1, 2, \ldots, b$.
44. $x_1^2, x_2^2, \ldots, x_b^2$ is a complete residue system modulo b.

C   45. Show by induction on the degree of P that if P is a polynomial with integer coefficients and if $x \equiv x' \pmod b$, then $P(x) \equiv P(x') \pmod b$.
46. Show that if $(a, b) = 1$ and if $x_1, x_2, \ldots, x_b$ is a complete residue system modulo b, then so is $ax_1, ax_2, \ldots, ax_b$.
47. Prove or disprove: If P is a polynomial such that $P(x)$ is an integer for all integers x, and if $x \equiv x' \pmod b$, then $P(x) \equiv P(x') \pmod b$.
48. Prove that if $d|b$ and if $t' \equiv t \pmod d$, then $x_0 + bt/d \equiv x_0 + bt'/d \pmod b$.

49. Suppose m objects are put into n boxes. Consider the following statements: (a) each box has at least one object; (b) no box has more than one object; (c) $m \geq n$; (d) $m \leq n$. Which pairs of statements imply the other two statements?

4.2   The Chinese Remainder Theorem

Simultaneous Congruences

Let us assume that a and b are relatively prime positive integers, and consider the expression $ax + by$ that proved useful in the last section. We have

$$ax + by \equiv ax + 0 \cdot y \equiv ax \quad (\text{mod } b),$$

and we know from Theorem 4.1 that by properly choosing x we can make ax fall into any desired congruence class (mod b). But in the same way

$$ax + by \equiv 0 \cdot x + by \equiv by \quad (\text{mod } a),$$

and by can be put into any desired congruence class modulo a by properly choosing y.

We have found a single integer (namely, $ax + by$) that is simultaneously in any two predetermined congruence classes modulo a and b, and have proved the following theorem.

THEOREM   Let a and b be relatively prime positive integers. Then for any integers k_1 and k_2, we can find an integer z such that

$$z \equiv k_1 \quad (\text{mod } a) \qquad \text{and} \qquad z \equiv k_2 \quad (\text{mod } b).$$

Example   Suppose we want to find an integer z such that

$$z \equiv 2 \quad (\text{mod } 5) \qquad \text{and} \qquad z \equiv -1 \quad (\text{mod } 6).$$

Here $a = 5$, $b = 6$, $k_1 = 2$, and $k_2 = -1$. A solution to

$$5x \equiv -1 \quad (\text{mod } 6)$$

is $x = 1$, and a solution to

$$6y \equiv 2 \quad (\text{mod } 5)$$

is $y = 2$. Thus the original set of congruences has the solution

$$z = ax + by = 5 \cdot 1 + 6 \cdot 2 = 17.$$

Note that indeed $17 \equiv 2 \pmod 5$ and $17 \equiv -1 \pmod 6$.

All Solutions to Two Congruences

Suppose z is one solution to

$$\left. \begin{array}{l} z \equiv k_1 \pmod a \\ z \equiv k_2 \pmod b \end{array} \right\}, \qquad (*)$$

where we are still assuming $(a, b) = 1$. We will try to decide what all solutions to the system are. First suppose z' is another solution. Then clearly

$$z' \equiv z \pmod a \qquad \text{and} \qquad z' \equiv z \pmod b.$$

This means that both a and b must divide $z - z'$. But since $(a, b) = 1$ the primes dividing a are all different from those dividing b. Thus we must have that ab divides $z' - z$ (see Theorem 2.2), or

$$z' \equiv z \pmod{ab}.$$

Since from Theorem 1.8 if $z' \equiv z \pmod{ab}$ then $z' \equiv z \pmod a$ and $z' \equiv z \pmod b$, we have proved the following theorem.

THEOREM 4.2 If $(a, b) = 1$, then $z' \equiv z \pmod a$ and $z' \equiv z \pmod b$ if and only if $z' \equiv z \pmod{ab}$.

Now if z is a solution to $(*)$ and if $z' \equiv z$ modulo both a and b, then z' is also a solution. Thus we have settled the question of what all solutions to $(*)$ are, given one.

THEOREM If $(a, b) = 1$ and if z is a solution to

$$z \equiv k_1 \pmod a \qquad \text{and} \qquad z \equiv k_2 \pmod b,$$

then z' is another solution if and only if $z' \equiv z \pmod{ab}$.

Example Professor Snabley feeds his pet python every four days and bathes it once a week. This week he fed it on Tuesday and washed it on Wednesday. When, if ever, will he feed and wash the python on the same day? How often will this happen?

Let us number the days, with this Tuesday as day 1. Then the snake will be fed on days 1, 5, 9, ..., and, in general, on day z exactly when

$z \equiv 1 \pmod 4$.

Since he washes the snake every seven days beginning with Wednesday (day 2), day z is a washday exactly when

$z \equiv 2 \pmod 7$.

We have $a = 4$, $b = 7$, $k_1 = 1$, and $k_2 = 2$. One solution to

$4x \equiv 2 \pmod 7$

is $x = 4$, and one solution to

$7y \equiv 1 \pmod 4$

is $y = 3$. Thus a solution to the simultaneous congruences is

$z = 4 \cdot 4 + 7 \cdot 3 = 37,$

and the snake will be washed and fed on day 37. In fact this will happen on any day congruent to 37 modulo $28 = 4 \cdot 7$. The next time it occurs will be on day $9 = 37 - 28$, which is Wednesday of next week.

A Formula for the Solution of Two Congruences

In the examples we have done so far the numbers were small enough that we could solve the congruences $ax \equiv k_2 \pmod b$ and $by \equiv k_1 \pmod a$ by inspection. If a and b were larger, however, we would use our Euclidean Algorithm method of solving these congruences. This would give us an x and y such that

$ax + by = 1 = (a, b)$

(we are still assuming a and b relatively prime), and so we would have

$ax \equiv 1 \pmod b$ and $by \equiv 1 \pmod a$.

Then $x' = k_2 x$ and $y' = k_1 y$ satisfy

$ax' \equiv k_2 \pmod b$ and $by' \equiv k_1 \pmod a$.

Thus $z = ak_2 x + bk_1 y$ provides a solution to $(*)$.

The advantages to calculating the solution to (∗) in this way are more than just the fact that the Euclidean Algorithm method yields the x and y at the same time, as we will see later.

We now summarize what we have figured out above in the following theorem, which gets its name from the fact that it was introduced into European mathematics from Chinese writings. It was known to the Chinese at least as early as the first century A.D.

THE CHINESE REMAINDER THEOREM   Consider the system of congruences

$$z \equiv k_1 \pmod{a}$$

$$z \equiv k_2 \pmod{b},$$

where $(a, b) = 1$. Let $ax \equiv 1 \pmod{b}$ and $by \equiv 1 \pmod{a}$. Then one solution to the system of congruences is

$$z = axk_2 + byk_1.$$

An integer z' is a solution if and only if $z' \equiv z \pmod{ab}$.

Example   Find the least positive solution to

$$z \equiv 15 \pmod{43}$$

$$z \equiv -8 \pmod{21}.$$

Here $a = 43$, $b = 21$, $k_1 = 15$, and $k_2 = -8$. Now

$$43 = 2 \cdot 21 + 1,$$

and so

$$1 \cdot 43 + (-2)21 = 1,$$

so we take $x = 1$ and $y = -2$. A solution to the original system is then

$$z = 43 \cdot 1(-8) + 21(-2)15 = -974.$$

In fact anything congruent to -974 modulo $903 = 43 \cdot 21$ is also a solution. The least positive solution is

$$-974 + 2 \cdot 903 = 832.$$

Example Find all solutions z, $0 < z < 100$, to

$$6z \equiv 3 \quad (\text{mod } 9)$$

$$3z \equiv 2 \quad (\text{mod } 8).$$

By the methods of the last section we find that a complete solution to the first congruence is $z = 2, 5, 8$ and a complete solution to the second congruence is $z = 6$. Thus the system is equivalent to

$$z \equiv 2, 5, \text{ or } 8 \quad (\text{mod } 9)$$

$$z \equiv 6 \quad (\text{mod } 8).$$

Although we could solve this as three separate problems, the formula of the Chinese Remainder Theorem allows most of the work to be done only once. We have $a = 9$ and $b = 8$. Then

$$9 \cdot 1 + 8(-1) = 1,$$

so we can take $x = 1$ and $y = -1$. We have $k_1 = 2, 5,$ or 8 and $k_2 = 6$, and so get the solutions

$$z = axk_2 + byk_1 = 9k_2 - 8k_1 = 54 - 8k_1.$$

Allowing k_1 to be 2, 5, and 8 yields $z = 38, 14,$ and -10.

Since anything congruent to one of these modulo $72 = 9 \cdot 8$ is also a solution, we also have in the range $0 < z < 100$ the solutions $14 + 72 = 86$, and $-10 + 72 = 62$. Thus the values of z satisfying the conditions of the problem are

$$z = 14, 38, 62, \text{ and } 86.$$

(The reason we can get more than 3 solutions is that $1, 2, \ldots, 99$ is not a complete residue system modulo 72.)

PROBLEMS FOR SECTION 4.2

A *For the problems in part A find all solutions of the system of congruences satisfying the inequality given.*

1. $z \equiv 3 \quad (\text{mod } 5)$
    $z \equiv 3 \quad (\text{mod } 8)$
    $0 \le z < 40$

2. $z \equiv -5 \quad (\text{mod } 6)$
    $z \equiv -5 \quad (\text{mod } 7)$
    $0 \le z < 42$

3. $z \equiv 3 \quad (\text{mod } 5)$
    $z \equiv 2 \quad (\text{mod } 8)$
    $0 \le z < 40$

4. $z \equiv 4 \quad (\text{mod } 6)$
    $z \equiv 5 \quad (\text{mod } 7)$
    $0 \le z < 42$

5. $z \equiv 20 \pmod 9$
   $z \equiv -3 \pmod{10}$
   $0 \le z < 90$

6. $z \equiv -2 \pmod 8$
   $z \equiv 4 \pmod{11}$
   $0 \le z < 88$

7. $x \equiv 2 \pmod 9$
   $x \equiv -2 \pmod{11}$
   $0 \le x < 150$

8. $x \equiv 8 \pmod{13}$
   $x \equiv 10 \pmod 5$
   $0 \le x < 50$

9. $t \equiv 5 \pmod{31}$
   $t \equiv 6 \pmod{33}$
   $0 \le t < 1023$

10. $t \equiv -3 \pmod{29}$
    $t \equiv 2 \pmod{16}$
    $0 \le t < 464$

11. $z \equiv 1$ or $3 \pmod 7$
    $z \equiv 4 \pmod 9$
    $0 \le z < 63$

12. $z \equiv 2 \pmod{11}$
    $z \equiv 5$ or $6 \pmod 8$
    $0 \le z < 88$

13. $w \equiv \pm 5 \pmod 7$
    $w \equiv 1$ or $2 \pmod{12}$
    $0 \le w < 84$

14. $w \equiv 3, 4,$ or $5 \pmod{10}$
    $w \equiv 2 \pmod 9$
    $0 \le w < 90$

15. $x \equiv 4 \pmod{13}$
    $x \equiv 1, 3,$ or $4 \pmod 8$
    $0 \le x < 132$

16. $x \equiv 1$ or $5 \pmod 9$
    $x \equiv 0$ or $\pm 1 \pmod{17}$
    $0 \le x < 123$

17. $2x \equiv 5 \pmod 7$
    $5x \equiv 1 \pmod 8$
    $0 \le x < 56$

18. $4x \equiv 17 \pmod 9$
    $-2x \equiv 5 \pmod{11}$
    $0 \le x < 99$

19. $6z \equiv 11 \pmod{19}$
    $7z \equiv 0 \pmod{20}$
    $0 \le z < 380$

20. $7z \equiv 8 \pmod{15}$
    $11z \equiv 13 \pmod 8$
    $0 \le z < 120$

21. Solve $x \equiv 2 \pmod 3$
    $x \equiv -1 \pmod 4$
    $x \equiv 3 \pmod 5$
    $0 \le x < 60$

22. Solve $x \equiv 5 \pmod 7$
    $x \equiv 1 \pmod 8$
    $x \equiv 1 \pmod 5$
    $0 \le x < 280.$

23. Give a proof that if $(a, b) = 1$, $a|m$, and $b|m$, then $ab|m$, based on the theorems of Sections 1.1 and 1.2.

24. Show that if $(a, b) = (a, c) = (b, c) = 1$, and if $a|m$, $b|m$, and $c|m$, then $abc|m$.

25. Show that if a, b, and c are positive integers with $(a, b) = (a, c) = (b, c) = 1$, then $z' \equiv z \pmod a$, $z' \equiv z \pmod b$, and $z' \equiv z \pmod c$ if and only if $z' \equiv z \pmod{abc}$.

26. Suppose $(a, b) = 1$. Show that as x runs through a complete residue system $\pmod b$ and y runs through a complete residue system $\pmod a$, then $ax + by$ runs through a complete residue system $\pmod{ab}$.

27. Professor Crittenden buys a new car every three years; he bought his first in 1981. He gets a sabbatical leave every seven years, starting in 1992. When will he first get both during a leap year?

28. Senator McKinley was first elected in 1962. His reelection is assured unless the campaign coincides with an attack of the seven-year-itch such as hit him in 1965. When must he worry first?

29. A certain baby demands to be fed every five hours. On Tuesday, July 22, its

mother watched the start of the NBC Today program at 7 A.M. while feeding the baby. When will this happen again?

30. A homeowner drains her hot water heater every 33 days and changes the filter in her heat pump every 56 days. She did the first on January 3, 1984 and the second on January 5, 1984. When after this does she first do both on the same day?

C 31. Show that if $(a, b) = (b, c) = (a, c) = 1$, and if k_1, k_2, and k_3 are any integers, then

$$x \equiv k_1 \pmod{a}$$

$$x \equiv k_2 \pmod{b}$$

$$x \equiv k_3 \pmod{c}$$

has a solution.

32. Show by induction on n that given integers $k_1, k_2, \ldots, k_n$ and positive integers $m_1, m_2, \ldots, m_n$ such that $(m_i, m_j) = 1$ for $i \neq j$, then

$$x \equiv k_1 \pmod{m_1}$$

$$x \equiv k_2 \pmod{m_2}$$

$$\vdots$$

$$x \equiv k_n \pmod{m_n}$$

is solvable.

33. Prove or disprove: If $(a, b) = (a, c) = (b, c) = 1$, and if x, y, and z run through complete residue systems modulo a, b, and c, respectively, then $ax + by + cz$ runs through a complete residue system modulo abc.

4.3 More Than Two Congruences

A More General Chinese Remainder Theorem

We now consider the problem of solving more than two congruences. Consider, for example, the system

$$z \equiv k_1 \pmod{a}$$

$$z \equiv k_2 \pmod{b}$$

$$z \equiv k_3 \pmod{c}.$$

We need an expression to fill the role that $ax + by$ did in the case of two congruences. The first thing thought of will probably be

$$ax + by + cw$$

(the letter z is already in use), but this doesn't work. The problem is that if we take this expression modulo a, b, and c, we find we want

$$by + cw \equiv k_1 \quad (\text{mod } a)$$

$$ax + cw \equiv k_2 \quad (\text{mod } b)$$

$$ax + by \equiv k_3 \quad (\text{mod } c),$$

and this system is more complicated than the one we started with.

A little experimentation leads us to the more useful expression

$$abx + bcy + acw.$$

This number will be a solution to the original system if

$$bcy \equiv k_1 \quad (\text{mod } a)$$

$$acw \equiv k_2 \quad (\text{mod } b)$$

$$abx \equiv k_3 \quad (\text{mod } c);$$

and if we assume that $(bc, a) = (ac, b) = (ab, c) = 1$, then by Theorem 4.1 all these congruences will be solvable. The conditions on a, b, and c are cumbersome as stated, but really only amount to the fact that no prime divides any two of a, b, and c.

We will give a name to the latter idea (for n rather than 3 numbers), and then proceed to give a version of the Chinese Remainder Theorem for n congruences. As in the version of the last section, we will multiply by the k's after solving congruences with 1 as the constant. The proof of the new version will be left for the problems at the end of the section.

DEFINITION   relatively prime in pairs
    We say the numbers $a_1, a_2, \ldots, a_n$ are **relatively prime in pairs** in case $(a_i, a_j) = 1$ whenever $i \neq j$. For example 3, 4, and 5 are relatively prime in pairs, but 4, 5, and 6 are not, even though no integer greater than 1 divides all three.

THE CHINESE REMAINDER THEOREM FOR n CONGRUENCES   Let $a_1, a_2, \ldots, a_n$ be positive integers, relatively prime in pairs. Consider the congruences

$$z \equiv k_1 \quad (\text{mod } a_1)$$

$$z \equiv k_2 \quad (\text{mod } a_2)$$

$$\vdots$$

$$z \equiv k_n \quad (\text{mod } a_n).$$

Let $A = a_1 a_2 \cdots a_n$. Let x_i be a solution to

$$\left(\frac{A}{a_i}\right) x_i \equiv 1 \quad (\text{mod } a_i),$$

for $i = 1, 2, \ldots, n$. Then a solution to the original system of congruences is

$$z = \sum_{i=1}^{n} k_i \left(\frac{A}{a_i}\right) x_i.$$

Furthermore, z' is another solution if and only if $z' \equiv z \,(\text{mod } A)$.

Example   Consider the system

$$z \equiv 2 \quad (\text{mod } 3)$$

$$z \equiv 5 \quad (\text{mod } 4)$$

$$z \equiv -3 \quad (\text{mod } 7).$$

This is solvable since 3, 4, and 7 are relatively prime in pairs. We have

$$a_1 = 3, \qquad a_2 = 4, \qquad a_3 = 7$$

$$k_1 = 2, \qquad k_2 = 5, \qquad k_3 = -3,$$

and $A = 3 \cdot 4 \cdot 7 = 84$. The congruences

$$\left(\frac{84}{3}\right) x_1 \equiv 28 x_1 \equiv x_1 \equiv 1 \quad (\text{mod } 3)$$

$$\left(\frac{84}{4}\right) x_2 \equiv 21 x_2 \equiv x_2 \equiv 1 \quad (\text{mod } 4)$$

$$\left(\frac{84}{7}\right) x_3 \equiv 12 x_3 \equiv 5 x_3 \equiv 1 \quad (\text{mod } 7)$$

have the solutions $x_1 = 1$, $x_2 = 1$, and $x_3 = 3$. Thus a solution to the original system is

$$z = 2 \cdot 28 \cdot 1 + 5 \cdot 21 \cdot 1 + (-3)12 \cdot 3 = 53.$$

(The reader should check that 53 does satisfy the original system of congruences.) An integer z' is a solution if and only if $z' \equiv 53 \pmod{84}$, so other solutions are $-31 \, (=53 - 84)$, $137 \, (=53 + 84)$, etc.

Example   Find all solutions z, $0 < z < 500$, to

$$z \equiv 1 \pmod 2$$

$$z \equiv 2 \pmod 3$$

$$z \equiv 3 \pmod 5$$

$$z \equiv 4 \pmod 7.$$

Here we must solve the congruences

$$105x_1 \equiv 1 \pmod 2$$

$$70x_2 \equiv 1 \pmod 3$$

$$42x_3 \equiv 1 \pmod 5$$

$$30x_4 \equiv 1 \pmod 7.$$

We can take $x_1 = 1$, $x_2 = 1$, $x_3 = 3$, and $x_4 = 4$. Thus a solution is

$$z = 1 \cdot 105 \cdot 1 + 2 \cdot 70 \cdot 1 + 3 \cdot 42 \cdot 3 + 4 \cdot 30 \cdot 4 = 1103.$$

Subtracting multiples of $2 \cdot 3 \cdot 5 \cdot 7 = 210$ from 1103 gives additional solutions 893, 683, 473, 263, 53, $-157, \ldots$. The solutions desired are $z = 53$, 263, and 473.

Reduced Residue Systems

In the course of the last few sections we have used the idea of a complete residue system as a tool to describe the solutions of congruences. Along the way certain facts about complete residue systems themselves have arisen that we will now state explicitly.

For example Theorem 4.1 says we can get a complete solution to

$$ax \equiv c \pmod b$$

from a particular one by considering $x = x_0 + bt/(a, b)$, where t runs through a complete residue system modulo (a, b). If $(a, b) = 1$, then only one value of t

is allowed, and so the congruence has exactly one solution in any complete residue system modulo b. Plugging in incongruent x's must produce incongruent numbers ax. This brings us to the following theorem.

THEOREM   If $(a, b) = 1$, then as x runs through a complete residue system modulo b, so does ax.

If both a and x are relatively prime to b, then so is ax, by Theorem 1.13. Thus restricting x to those integers in a complete residue system relatively prime to b similarly restricts ax. We introduce some new language so we can express this more gracefully.

DEFINITION   reduced residue system

By a **reduced residue system modulo** b we mean the set of all integers relatively prime to b in any complete residue system modulo b.

Example   Since 1, 2, 3, 4, 5, 6 is a complete residue system modulo 6, a reduced residue system modulo 6 consists of 1 and 5. Others are $\{7, 5\}$, $\{-1, 1\}$, and $\{61, 65\}$. One reduced residue system modulo 5 is $\{1, 2, 3, 4\}$; another is $\{1, 12, 23, 34\}$.

Note that any two reduced residue systems modulo b have the same number of elements, namely, the number of congruence classes of elements relatively prime to b. Since the number of integers in the complete residue system 1, 2, . . . , b relatively prime to b is by definition $\varphi(b)$:

Any reduced residue system modulo b has $\varphi(b)$ elements.

We now have an improved version of the last theorem (see the argument before the definition), which we state as follows:

THEOREM 4.3   If $(a, b) = 1$, then as x runs through a complete or reduced residue system modulo b, so does ax.

Example   A complete residue system modulo $b = 8$ is $\{-2, -1, 0, 1, 2, 3, 4, 5\}$, and a reduced residue system is $\{-1, 1, 3, 5\}$. We take $a = 3$. The reader should check that $\{-6, -3, 0, 3, 6, 9, 12, 15\}$ is also a complete residue system (mod 8), and that $\{-3, 3, 9, 15\}$ is a reduced residue system (mod 8).

PROBLEMS FOR SECTION 4.3

Solve the systems of the first ten problems.

1. $x \equiv 2$ (mod 3)
   $x \equiv 2$ (mod 5)
   $x \equiv 2$ (mod 7)
   $0 < x < 200$

2. $z \equiv 3$ (mod 4)
   $z \equiv 3$ (mod 5)
   $z \equiv 3$ (mod 7)
   $0 < z < 200$

3. $x \equiv 1 \pmod 5$
   $x \equiv -2 \pmod 7$
   $x \equiv 7 \pmod 2$
   $0 \le x < 70$

4. $t \equiv 2 \pmod 3$
   $t \equiv 1 \pmod 4$
   $t \equiv -1 \pmod 5$
   $0 \le t < 60$

5. $z \equiv 0 \pmod 9$
   $z \equiv 3 \pmod 2$
   $z \equiv 10 \pmod 5$
   $0 \le z < 90$

6. $z \equiv 8 \pmod 4$
   $z \equiv 4 \pmod 3$
   $z \equiv 1 \pmod{11}$
   $0 \le z < 132$

7. $z \equiv 3 \pmod 2$
   $z \equiv 1 \pmod 3$
   $z \equiv -1 \pmod 5$
   $z \equiv -4 \pmod 7$
   $0 \le z < 210$

8. $q \equiv 1 \pmod 3$
   $q \equiv 2 \pmod 4$
   $q \equiv -5 \pmod 5$
   $q \equiv 5 \pmod 7$
   $0 \le q < 420$

9. $q \equiv 3 \pmod 4$
   $q \equiv 5 \pmod 7$
   $q \equiv -3 \pmod{11}$
   $0 \le q < 502$

10. $e \equiv 3 \pmod 8$
    $e \equiv 6 \pmod 7$
    $e \equiv 2 \pmod 5$
    $0 \le e < 316$

11. Which of the following are reduced residue systems (mod 5)?
    (a) $-1, 0, 1, 2$ (b) $-2, 0, 2, 4$ (c) $2, 4, 6, 8$
    (d) $4, 6, -4, 2$ (e) $7, 14, 21$ (f) $2, 4, 6, 8, 12$

B 12. Let $p, q,$ and r be distinct primes. Show there exists n such that $p|n$, $q|n + 1$, and $r|n + 2$.

13. Solve $x \equiv 2 \pmod 8$
    $x \equiv 1 \pmod 7$
    $x \equiv 3 \pmod 6$
    $0 \le x < 336.$

14. Solve $3x \equiv 9 \pmod 6$
    $2x \equiv 1 \pmod 5$
    $x \equiv 2 \pmod 7$
    $0 \le x < 210.$

The next four problems complete the proof of the Chinese Remainder Theorem for n congruences. Assume the notation and assumptions of the theorem in them.

15. Show the congruence $(A/a_i)x_i \equiv 1 \pmod{a_i}$ is solvable for $i = 1, 2, \ldots, n$.
16. Show the number z given in the theorem satisfies the original system of congruences.
17. Show that if z is a solution to the system and if $z' \equiv z \pmod A$, then z' is also a solution.
18. Show that if z and z' are solutions to the system of congruences, then $z' \equiv z \pmod A$.

True-False. In the next six problems tell which statements are true, and give counterexamples for those that are false. Assume that $x_1, x_2, \ldots, x_t$ is a reduced residue system (mod b), and that $(a, b) = 1$.

19. $2x_1, 2x_2, \ldots, 2x_t$ is a reduced residue system $\pmod b$.
20. $x_1 + 2, x_2 + 2, \ldots, x_t + 2$ is a reduced residue system $\pmod b$.

21. $x_1 + a, x_2 + a, \ldots, x_t + a$ is a reduced residue system (mod b).
22. $x_1^2, x_2^2, \ldots, x_t^2$ is a reduced residue system (mod b).
23. $-x_1, -x_2, \ldots, -x_t$ is a reduced residue system (mod b).
24. $b = \varphi(t)$.

25. On the Fourth of July Sam took a red pill at 8 A.M. and a green pill at noon. The next day he took a white pill at 10 P.M. He takes the red, green, and white pills every 5, 7, and 24 hours, respectively. How often does he take all three together? When in August will this first happen, if at all?

26. Leroy gave each of his four children the same amount of money. One spent all but 13 cents on 5-cent candy bars. (This is a very old problem.) The second spent all but 3 cents on 6-cent popsicles. The third spent all but 2 cents on 11-cent comic books. The fourth bought a $3 game but didn't have enough money to buy another. How much money did each child get?

27. Suppose a, b, and c are relatively prime in pairs. Show that if $(abx + bcy + acz, abc) = 1$, then $(x, c) = (y, a) = (z, b) = 1$.

28. Mr. Brown washes his car every 8 days, checks the oil every 7 days, and checks the tire pressure every 9 days. He washed it Monday, November 23, 1981, and checked the oil and tires Wednesday and Thursday of that week, respectively. When is the first day after that Monday on which he does all three things?

29. Give a set of positive integers that is a reduced residue system both modulo 5 and modulo 8.

30. Seven children found some jelly beans. When they divided them as evenly as possible among themselves, 1 was left over. They gave it to the cat, who ate it. Then the sister of one of the children appeared, so they redivided the jelly beans among the 8 children. This time there were 6 left over, which the cat ate. Then another child appeared. When they divided the jelly beans among the 9 children, 6 were left over. What is the least number of jelly beans they can have started with?

C 31. Suppose $(a, b) = 1$. Show that as x runs through a reduced residue system (mod b) and y runs through a reduced residue system (mod a), then $ax + by$ runs through a reduced residue system (mod ab).

32. Use the previous problem to give a proof that the Euler phi-function is multiplicative that does not depend on the formula derived for it in Section 3.5.

In the next three problems suppose that a, b, and c are positive integers, relatively prime in pairs.

33. Show that $(x, c) = (y, a) = (z, b) = 1$ implies $(abx + bcy + acz, abc) = 1$.

34. Show that as x, y, and z run through complete residue systems modulo c, a, and b, respectively, then $abx + bcy + acz$ runs through a complete residue system (mod abc).

35. Do the last problem with the word "complete" replaced by "reduced."

36. An integer n is said to be **square free** in case there is no prime p such that $p^2 \mid n$. Show that if $k > 0$ there exist k consecutive integers that are not square free.

4.4 The Theorems of Fermat and Euler

How Powers Fall into Congruence Classes

Assume a and b are relatively prime. Let us examine the powers of a (mod b). Taking $a = 2$ and $b = 7$, for example, we have

n	1	2	3	4	5	6	7
a^n	2	4	$8 \equiv 1$	$16 \equiv 2$	$32 \equiv 4$	$64 \equiv 1$	$128 \equiv 2$

Here all the congruences are modulo 7. We have done more work in the above table than we needed to, since for instance once we knew that $2^4 \equiv 2$ (mod 7), we had

$$2^5 \equiv 2 \cdot 2^4 \equiv 2 \cdot 2 = 4 \quad (\text{mod } 7).$$

From this point of view it is clear that the pattern of least residues 2, 4, 1, 2, 4, 1, ... will repeat forever. In the general case, since there are only finitely many congruence classes (mod b), eventually two powers of a will have to fall into the same class. But if

$$a^i \equiv a^j \quad (\text{mod } b), \qquad\qquad (*)$$

then

$$a^{i+1} \equiv a^{j+1} \quad (\text{mod } b)$$

follows from multiplication by a; and the pattern is repeated.

Let us look at this process more closely. Suppose (*) holds with $i < j$. This can be written

$$a^i \cdot 1 \equiv a^i a^{j-i} \quad (\text{mod } b).$$

Now $(a^i, b) = 1$ by Theorem 1.13, and so by the Cancellation Theorem

$$a^{j-i} \equiv 1 \quad (\text{mod } b).$$

We see that some power of a must be congruent to 1 (mod b).

DEFINITION the order of a (mod b)

Let $(a, b) = 1$. By the **order of** a (**mod** b) we mean the least positive integer k such that

$$a^k \equiv 1 \pmod{b}.$$

Examples From our previous calculation the order of 2 (mod 7) is 3. The following table shows the order of 1 through 6 (mod 7).

n	1	2	3	4	5	6	order
a							
1	1						1
2	2	4	1				3
3	3	2	6	4	5	1	6
4	4	2	1				3
5	5	4	6	2	3	1	6
6	6	1					2

In the table the powers have been replaced by congruent elements of the reduced residue system $1, 2, \ldots, 6$; for example $3^2 = 9 \equiv 2$ (mod 7).

Suppose a has the order k (mod b). Then if m is any multiple of k, say $m = kt$, we have

$$a^m = (a^k)^t \equiv 1^t \equiv 1 \pmod{b}.$$

Conversely, if $a^n \equiv 1$ (mod b), use the Division Algorithm to write

$$n = kq + r, \quad 0 \le r < k.$$

Then

$$a^n = a^{kq+r} = (a^k)^q a^r \equiv a^r \equiv 1 \pmod{b}.$$

But this contradicts the assumption that k is the smallest positive integer such that $a^k \equiv 1$ (mod b) unless $r = 0$. Thus we see that $k \mid n$.

We have already seen that if $a^i \equiv a^j$ (mod b) with $i < j$, then

$$a^{j-i} \equiv 1 \pmod{b},$$

and so $k \mid j - i$. In particular no two of the powers of a up to the kth power can be congruent (mod b).

We sum up what we have proved in the following theorem.

THEOREM 4.4 Let $(a, b) = 1$, and let k be the order of a (mod b). Then the numbers

$$a^1, a^2, a^3, \ldots, a^k$$

are incongruent (mod b) and every positive power of a is congruent to one of them. Furthermore $a^m \equiv 1$ (mod b) if and only if $k \mid m$.

Euler's Theorem

We found that the 6 elements of a reduced residue system (mod 7) had orders as follows:

element	1	2	3	4	5	6
order	1	3	6	3	6	2

Notice that the orders, 1, 2, 3, and 6, are all divisors of $6 = \varphi(7)$.

If we take $b = 15$, then a reduced residue system has $\varphi(15) = 8$ elements. The reader should confirm that the following table gives the correct orders of these elements.

element	1	2	4	7	8	11	13	14
order	1	4	2	4	4	2	4	2

Here the orders are 1, 2, and 4, all divisors of 8. In each case every order is a divisor of $\varphi(b)$. It appears that it may be true that the order of any element a (mod b) is always a divisor of $\varphi(b)$. In light of the last theorem this is equivalent to saying that

$$a^{\varphi(b)} \equiv 1 \quad (\text{mod } b)$$

whenever $(a, b) = 1$.

In order to prove this it seems natural to try to associate the number a with a reduced residue system (mod b) in some way, since the latter has exactly $\varphi(b)$ elements. Theorem 4.3 suggests multiplication; if $x_1, x_2, \ldots, x_t$ form a reduced residue system (mod b) (where $t = \varphi(b)$), then so do $ax_1, ax_2, \ldots, ax_t$.

In particular, the numbers $ax_1, ax_2, \ldots, ax_t$ are congruent in some order to the numbers $x_1, x_2, \ldots, x_t$. We do not know exactly how these elements match up, but certainly if we multiply all the congruences together and sort things out we get

$$ax_1 ax_2 \cdots ax_t \equiv x_1 x_2 \cdots x_t \quad (\text{mod } b),$$

or

$$a^{\varphi(b)}(x_1 x_2 \cdots x_t) \equiv x_1 x_2 \cdots x_t \quad (\text{mod } b).$$

By Theorem 1.13 the product of the x's is relatively prime to b and so the Cancellation Theorem yields

$$a^{\varphi(b)} \equiv 1 \quad (\text{mod } b).$$

EULER'S THEOREM If b is a positive integer and if $(a, b) = 1$, then

$$a^{\varphi(b)} \equiv 1 \pmod{b}.$$

Combining Euler's Theorem with the last line of Theorem 4.4 we get the following result.

THEOREM 4.5 If $(a, b) = 1$ and if k is the order of $a \pmod{b}$, then $k \mid \varphi(b)$.

Applications of Euler's Theorem

Let us use Euler's Theorem to find the last digit of 17^{100}. What we are after is the least residue of $17^{100} \pmod{10}$. Now $\varphi(10) = 4$, so by Euler's Theorem

$$17^4 \equiv 1 \pmod{10}.$$

Since $4 \mid 100$, $17^{100} \equiv 1 \pmod{10}$ also; and the answer is 1.

Suppose we want the last *two* digits of 17^{100}. We want the least residue of $17^{100} \pmod{100}$. Since $\varphi(100) = 40$, Euler's Theorem tells us that

$$17^{40} \equiv 1 \pmod{100}.$$

Now $100 = 2 \cdot 40 + 20$, so

$$17^{100} \equiv (17^{40})^2 17^{20} \equiv 17^{20} \pmod{100},$$

and we have reduced the exponent to 20.

This still leaves us with an unpleasant computation, however. It turns out that there is a way to break down the problem to make it more manageable. By Theorem 4.2 the congruence

$$z \equiv 17^{20} \pmod{100}$$

is equivalent to the system of congruences

$$z \equiv 17^{20} \pmod{4}$$

$$z \equiv 17^{20} \pmod{25}.$$

We will solve these one at a time. Since $\varphi(4) = 2$, we have

$$17^2 \equiv 1 \pmod{4},$$

and so $17^{20} \equiv 1 \pmod{4}$ also. Thus the first congruence is equivalent to

$$z \equiv 1 \pmod{4}.$$

Likewise $\varphi(25) = 20$, and so

$$17^{20} \equiv 1 \quad \text{(mod 25)}.$$

We see that the second congruence is equivalent to

$$z \equiv 1 \quad \text{(mod 25)}.$$

Thus we have reduced our problem to solving the system

$$z \equiv 1 \quad \text{(mod 4)}$$

$$z \equiv 1 \quad \text{(mod 25)}.$$

In this case an obvious solution is $z = 1$, and every $z \equiv 1$ (mod 100) is a solution. (If we hadn't been so lucky we would have had to solve a Chinese Remainder Theorem problem at this point.) We see that the least residue of 17^{100} (mod 100) is 1, and so its last two digits are 0 and 1.

Note that in order to apply Theorem 4.2 in this example we required the moduli 4 and 25 to be relatively prime.

Fermat's Theorem

The case of Euler's Theorem when the modulus b is a prime p is attributed to Fermat. Actually Fermat merely told people he had proved the theorem bearing his name; Euler first published a proof. Since $\varphi(p) = p - 1$ we have the following theorem.

FERMAT'S THEOREM   If p is prime and if $p \nmid a$, then

$$a^{p-1} \equiv 1 \quad \text{(mod } p\text{)}.$$

Of course since p is prime $(a, p) = 1$ is equivalent to $p \nmid a$. A slight variation allows this theorem to be stated with a simpler hypothesis.

FERMAT'S THEOREM (second form)   If p is prime, then

$$a^p \equiv a \quad \text{(mod } p\text{)}$$

for all integers a.

Proof   If $p \nmid a$, then multiplying the congruence of the first form of the theorem by a gives the conclusion. But if $p \mid a$, then both sides of the new congruence are congruent to 0 (mod p).

Wilson's Theorem

The argument used to prove Euler's Theorem above is too good to let alone. Recall that we derived the congruence

$$x_1 x_2 \cdots x_t \equiv a x_1 a x_2 \cdots a x_t \pmod{b},$$

where $(a, b) = 1$, from the fact that if $x_1, x_2, \ldots, x_t$ was a reduced residue system (mod b) then so was $a x_1, a x_2, \ldots, a x_t$.

Matching things up is the germ of this proof, i.e., matching elements x with congruent elements ax. Let us try to evaluate (mod b) the product $x_1 x_2 \cdots x_t$ itself by a similar argument.

We know that for each x in a reduced residue system (mod b) there exists a unique element x' in the system such that $x x' \equiv 1 \pmod{b}$. Our idea will be to pair up the x's this way, each pair multiplying to 1 (mod b).

We seem to have proved that $x_1 x_2 \cdots x_t \equiv 1 \pmod{b}$, but a little care is needed. We must consider the possibility that x pairs with itself, that is $xx \equiv 1 \pmod{b}$. This is certainly the case with $x \equiv \pm 1 \pmod{b}$, and perhaps for other x's; for example $4 \cdot 4 \equiv 1 \pmod{15}$.

If we assume the modulus is a prime p we simplify the situation, since if

$$x^2 \equiv 1 \pmod{p},$$

then $p \mid x^2 - 1 = (x - 1)(x + 1)$. Theorem 2.1 says that p divides one or the other of these factors, and so

$$x \equiv \pm 1 \pmod{p}.$$

A prime also has the advantage of a very explicit reduced residue system, namely $1, 2, \ldots, p - 1$.

If $p = 11$, for example, we can pair off everything except 1 and 10 ($\equiv -1$). Indeed

$$2 \cdot 6 \equiv 3 \cdot 4 \equiv 5 \cdot 9 \equiv 7 \cdot 8 \equiv 1 \pmod{11}.$$

Thus

$$1 \cdot 2 \cdot 3 \cdot 4 \cdot 5 \cdot 6 \cdot 7 \cdot 8 \cdot 9 \cdot 10 = 1(2 \cdot 6)(3 \cdot 4)(5 \cdot 9)(7 \cdot 8)10$$

$$\equiv -1 \pmod{11}.$$

The same argument works for any odd prime p, producing a theorem that bears John Wilson's name, even though there is no evidence that Wilson (1741–1793) did more than guess it from numerical evidence. The first published proof was by Lagrange.

WILSON'S THEOREM   If p is prime, then

$$(p - 1)! \equiv -1 \pmod{p}.$$

The statement (without proof) of this theorem was first published by the English mathematician Edward Waring in 1770 in his *Meditationes Algebraicae*, along with two other famous conjectures based on numerical evidence. One of these, known as **Waring's Problem**, says that each positive integer is the sum of at most 4 squares, of at most 9 cubes, etc. That is, given a positive integer k there exists a least integer $g(k)$ such that each positive integer is the sum of at most $g(k)$ kth powers. (The first proof that $g(2) = 4$ was given by Lagrange.) Waring's Problem was not settled in general until 1909, when the German mathematician David Hilbert proved it. (Hilbert did not give a formula for $g(k)$, but merely showed it always existed.)

The other conjecture, due to Christian Goldbach, says that each even integer greater than 2 can be written as the sum of two primes. This still unproved **Goldbach Conjecture** was discussed in Chapter 0.

PROBLEMS FOR SECTION 4.4

A *In the first four problems find the order of each element in the given reduced residue system M' (mod m).*

1. $m = 5$, $M' = \{1, 2, 3, 4\}$
3. $m = 11$, $M' = \{1, 2, \ldots, 10\}$

2. $m = 10$, $M' = \{1, 3, 7, 9\}$
4. $m = 13$, $M' = \{1, 2, \ldots, 12\}$

In the next six problems find the order of a (mod b). Compute $\varphi(b)$.

5. $a = 3$, $b = 16$
8. $a = 18$, $b = 25$

6. $a = 5$, $b = 17$
9. $a = 10$, $b = 19$

7. $a = 25$, $b = 18$
10. $a = 10$, $b = 21$

In the next six problems compute the least residue of a^k (mod b).

11. $a = 2$, $k = 35$, $b = 11$
13. $a = 50$, $k = 40$, $b = 19$
15. $a = 8$, $k = 50$, $b = 21$

12. $a = 3$, $k = 45$, $b = 13$
14. $a = 60$, $k = 70$, $b = 19$
16. $a = 9$, $k = 43$, $b = 25$

17. Find the last two digits of 7^{77}.   18. Find the last two digits of 9^{99}.

In each of the next four problems find the least residue of $(b - 1)!$ (mod b).

19. $b = 9$ 20. $b = 10$ 21. $b = 101$ 22. $b = 21$

B 23. Prove or disprove: If b is composite, then $(b - 1)! \equiv 0 \pmod{b}$.

In the next four problems find the least residue of a^k (mod b). Use the method following Euler's Theorem.

24. $a = 35$, $k = 10$, $b = 72$
26. $a = 10$, $k = 75$, $b = 99$

25. $a = 33$, $k = 10$, $b = 56$
27. $a = 100$, $k = 1000$, $b = 91$

28. Show that if $b > 1$ is not prime, then $(b - 1)! \not\equiv -1 \pmod{b}$.
29. Show that if $(a, b) = 1$, then $a^i \equiv a^j \pmod{b}$ if and only if $i \equiv j \pmod{k}$, where k is the order of $a \pmod{b}$.
30. Give an example of integers k and b such that $k \mid \varphi(b)$ and yet no integer has order $k \pmod{b}$.
31. Show that if $(a, b) = 1$ and if $a^* \equiv a^k \pmod{b}$ for some positive integer k, then the order of $a^* \pmod{b}$ does not exceed the order of $a \pmod{b}$.
32. Show that no integer has order 40 (mod 100).
33. Show that no integer has order 24 (mod 72).
34. Show that if $x^2 \equiv y^2 \pmod{p}$, p a prime, then $x \equiv \pm y \pmod{p}$.
35. Illustrate the proof of Wilson's Theorem for $p = 13$ by pairing off the integers $1, 2, \ldots, 12$.
36. Show how the proof given for Wilson's Theorem breaks down for $p = 2$.
37. Show that if $(a, b) = 1$, then one solution to $ax \equiv c \pmod{b}$ is $x = ca^{\varphi(b)-1}$.
38. Show that if $b > 2$, then there exists an integer of order 2 (mod b).
39. Show that $\varphi(n)$ is even for $n > 2$.
40. Prove that $g(2) > 3$, where g is as defined at the end of this section.
41. Show that if p is prime, then $(p - 2)! \equiv 1 \pmod{p}$.
42. Show that if p is an odd prime, then $(p - 3)! \equiv (p - 1)/2 \pmod{p}$.
43. Show that if p is an odd prime and $p \nmid a$, then $a^{(p-1)/2} \equiv \pm 1 \pmod{p}$.
44. We call p a **twin prime** if p is prime and also $p - 2$ or $p + 2$ is prime. Find the smallest even integer greater than 4 that is not the sum of two twin primes. (The two primes do not have to be twins of each other.)

<u>C</u> *In the next two problems assume a has order k (mod p), p a prime. Assume $d > 0$.*

45. Show that if $(d, k) = 1$, then a^d has order $k \pmod{p}$.
46. Show that if $(d, k) > 1$, then a^d has order $< k \pmod{p}$.
47. Suppose that a has order $p - 1 \pmod{p}$, p prime. Show that the order of $a^j \pmod{p}$ is $(p - 1)/(j, p - 1)$.
48. Assume the hypotheses of the last problem. Show that if $k \mid p - 1$, then any reduced residue system (mod p) contains exactly $\varphi(k)$ elements of order k (mod p).
49. Prove Fermat's Theorem (second version) for positive integers a by induction on a, using the Binomial Theorem. (A statement of the Binomial Theorem may be found in the exercises for Section 5.2.)

4.5 Public-Key Cryptography

The Idea of a Key

Cryptography is a method of sending a message in such a form that only its intended recipient can understand it. Although this calls to mind spies, diplomats, and the military, cryptography has a wider application. In an age when more and more information is transmitted over telephone lines or by radio (and so is subject to interception), keeping one's messages secret is an increasing problem for businesses and individuals.

Like everything else, cryptography has its own special language. The original message is called the **plaintext**, and the (supposedly) unreadable version of it is called the **ciphertext**. The processes of going from plaintext to ciphertext and back are called **enciphering** and **deciphering**, respectively.

Often enciphering methods involve a **key** that is known to the sender and intended receiver of the message but no one else. The idea is that someone who does not know the key will not be able to decipher the message even if he knows the general method of encipherment.

We will illustrate the idea of a key with a substitution cipher, one of the oldest and simplest methods. Substitution amounts to merely replacing each letter of the alphabet with another letter. Although any message of a few sentences or more enciphered by substitution may be easily figured out (in fact such problems appear in newspapers and crossword puzzle magazines), our aim with this example is simply to show the concepts involved.

We will use as our key the words "number theory." First we cross out any repeated letters, leaving

    NUMBERTHOY.

Now we write these letters under the alphabet, followed by the unused letters of the alphabet, in order.

    Plaintext: ABCDEFGHIJKLMNOPQRSTUVWXYZ
    Ciphertext: NUMBERTHOYACDFGIJKLPQSVWXZ

To encipher we merely replace each letter of our message with the corresponding ciphertext letter. Suppose our message is SEND MONEY. We would replace the S by L, since L is below S in our table, etc. We get

    Plaintext: SEND MONEY
    Ciphertext: LEFB DGFEX.

The message we would send is LEFB DGFEX. Even though substitution is inadequate to assure the security of a longer message, it is doubtful that anyone could decipher LEFB DGFEX, since the repeated letters provide the only clues. The message could be anything with the same pattern of repeated letters, for example LIST MUSIC, MINE TUNIS, or DRAW CHART.

Of course to decipher LEFB DGFEX the intended receiver of the message would use the key to make his own table like the one above. He would then use it in reverse, changing L to S, E to E, F to N, etc., to retrieve the original message.

The RSA Method

In any method of encipherment using a secret key, the key itself is a weak link. It must be known to both parties who want to communicate, so a breach in security by either will destroy the secrecy of messages. If the key is so complicated that it must be written down, opportunities for discovering it increase. Also it must be prearranged. Parties wishing to communicate secretly for the first time must first agree on the key by some more secure method, such as a courier.

These reasons have led to the development of **public-key cryptography**. Such a system uses a key which allows encipherment but not decipherment of messages. Each party then publishes his own key for anyone who wants to communicate with him to use. He keeps secret additional information which allows him alone to decipher incoming messages.

Systems as we have just described may seem impossible at first, since in traditional enciphering methods knowing how to encipher and decipher go hand in hand. To be accurate, we must admit that in the public-key systems decipherment by unauthorized parties cannot be said to be impossible. The goal is merely to place such decipherment beyond available resources of time and money. If it would take the world's largest computer 1000 years using the most efficient methods known to decipher a message, then the message may be considered secure, at least unless better methods are found.

In actual use the public-key methods we will describe would be implemented using computers, both for encipherment and decipherment. Large numbers would be used to make unauthorized decipherment, even using computers, impractical. In this section, however, we will scale the systems down to human dimensions, so that encipherment and decipherment are reasonable tasks for a person (a hand calculator will make them even easier).

The method we will describe is called the **RSA method**. The name is taken from the initials of its inventors, R.L. Rivest, A. Shamir, and L. Adleman, of the Massachusetts Institute of Technology. The security of messages enciphered by the method depends on the difficulty of factoring a large number n. This number will be the product of two large primes, p and q. To be safe from unauthorized decipherment by use of large computers, p and q should each be about 100 (decimal) digits long, making $n = pq$ a number of about 200 digits. To illustrate the method, however, we will use much smaller numbers, say $p = 29$ and $q = 41$. Then their product is $n = 1189$. We also need a positive integer e that is relatively prime to $\varphi(n) = (p - 1)(q - 1)$. In our example $\varphi(n) = 28 \cdot 40$, and we will take $e = 3$.

Now we publish the numbers n and e, and this pair constitutes our public key. In the example our public key is the pair $n = 1189$, $e = 3$. We keep the numbers p and q secret, not even revealing them to those from whom we expect secret messages.

RSA Enciphering

Suppose someone wants to send us a message no one else can read. This person first converts the message to numerical form, and breaks the resulting sequence of digits into blocks having fewer digits than n. In our example we will use the following simple conversion.

symbol	number	symbol	number	symbol	number	symbol	number
space	00	G	07	N	14	U	21
A	01	H	08	O	15	V	22
B	02	I	09	P	16	W	23
C	03	J	10	Q	17	X	24
D	04	K	11	R	18	Y	25
E	05	L	12	S	19	Z	26
F	06	M	13	T	20		

Suppose the message to be sent is SEND MONEY. Then the corresponding sequence of digits is:

19 05 14 04 00 13 15 14 05 25.

Since $n = 1189$ has four digits, our correspondent will break the message into groups of three digits:

190 514 040 013 151 405 250.

(An extra 0 is added to make the message come out even.)

We label these integers $P_1, P_2, \ldots$. The "P" stands for plaintext; since this method of transforming the message into a sequence of numbers is agreed upon beforehand and public knowledge, anyone could decipher the sequence at this stage.

The enciphered message is now the sequence $C_1, C_2, \ldots$, where C_i is defined to be the least residue of P_i^e (mod n). In our example C_i is the least residue of P_i^3 (mod 1189). Since the numbers n and e have been published, anyone who wants to correspond with us can encipher a message this way. (In practice, since the numbers involved would be very large, the encipherment would be done by a computer.)

In our example P_1 is 190, and so

$$C_1 \equiv 190^3 = 6859000 \equiv 848 \quad (\text{mod } 1189).$$

Thus we see $C_1 = 848$. (A method for finding least residues with a hand calculator is given at the end of Section 1.2.) Likewise we may compute C_2 by

$$C_2 \equiv 514^3 = 135796744 \equiv 1054 \quad (\text{mod } 1189).$$

If 514^3 is too big for our correspondent's calculator to handle he can compute its least residue in steps as follows.

$$514^3 = 514^2 514 = 264196 \cdot 514 \equiv 238 \cdot 514 = 122332 \equiv 1054 \quad (\text{mod } 1189).$$

In fact this calculation presages the method we will be forced to use soon when we need to raise numbers to powers larger than 3.

Continuing in this way he finds $C_1, \ldots, C_7$ to be

848, 1054, 983, 1008, 796, 695, 351,

and this is the ciphertext he sends to us.

RSA Deciphering

Having received the above message $C_1, C_2, \ldots$, now it is up to us to decipher it. We must undo the scrambling of the original message caused by raising the numbers $P_1, P_2, \ldots$ to the power e. This we will do by raising each number C_i to the power d, where d will be defined shortly. First we define the integer b by

$$b = [p - 1, q - 1].$$

Note that someone not knowing p and q will not be able to compute b. Now we define d to be the least positive solution to

$$ex \equiv 1 \pmod{b}.$$

In our example $b = [28, 40] = 28 \cdot 40/(28, 40) = 28 \cdot 40/4 = 280$, where the second equality follows from the theorem at the end of Section 1.2. Thus the congruence we must solve to find d is

$$3x \equiv 1 \pmod{280}.$$

Using the method of Section 4.1 we easily find the least positive solution to this to be $x = 187$, so this is our value of d.

Now all we have to do to retrieve the plaintext message is find the least residue of each ciphertext number C_i when raised to the power d. For example P_1 is just the least residue modulo 1189 of

$$C_1^d = 848^{187}.$$

Putting aside for the moment the question of *why* this will produce P_1, *how* are we going to do this? The number we intend to compute the least residue of is far too big for us to handle, even with the best calculator. The answer is, a little at a time.

Computing Big Powers

To compute numbers to the power d, we first convert d to its base 2, or binary, representation. One easy way to do this is to successively divide d by 2, keeping track of the remainders (which are all 0 or 1). For $d = 187$, for example, we have

$$187 = 93 \cdot 2 + 1$$

$$93 = 46 \cdot 2 + 1$$

$$46 = 23 \cdot 2 + 0$$

$$23 = 11 \cdot 2 + 1$$

$$11 = 5 \cdot 2 + 1$$

$$5 = 2 \cdot 2 + 1$$

$$2 = 1 \cdot 2 + 0$$

$$1 = 0 \cdot 2 + 1.$$

Then the remainders, listed in reverse order, give the binary representation of d. In our example

$$d = 187 = 10111011_2$$

$$= 1 \cdot 2^7 + 0 \cdot 2^6 + 1 \cdot 2^5 + 1 \cdot 2^4 + 1 \cdot 2^3 + 0 \cdot 2^2 + 1 \cdot 2 + 1$$

$$= 128 + 32 + 16 + 8 + 2 + 1.$$

Now in order to compute C_i to the power $d \pmod{n}$ we successively square it and reduce modulo n as follows.

k	$C^k \pmod{n}$
1	848
2	$848^2 = 719104 \equiv 948 \pmod{1189}$
4	$948^2 = 898704 \equiv 1009 \pmod{1189}$
8	$1009^2 = 1018081 \equiv 297 \pmod{1189}$
16	$297^2 = 88209 \equiv 223 \pmod{1189}$
32	$223^2 = 49729 \equiv 980 \pmod{1189}$
64	$980^2 = 960400 \equiv 877 \pmod{1189}$
128	$877^2 = 769129 \equiv 1035 \pmod{1189}.$

Then we have

$$848^{187} = 848^{128+32+16+8+2+1}$$

$$= 848^{128}848^{32}848^{16}848^{8}848^{2}848^{1}$$

$$\equiv 1035 \cdot 980 \cdot 223 \cdot 297 \cdot 948 \cdot 848 \pmod{1189}.$$

By multiplying out the last expression factor by factor, reducing modulo 1189 as we go, we find

$$848^{187} \equiv 190 \pmod{1189}.$$

Note that 190 is the original plaintext number P_1. In the same way we can calculate the other plaintext numbers 514, 40, 13, 151, 405, and 250. We list these as 3-place numbers,

190 514 040 013 151 405 250,

regroup into pairs,

19 05 14 04 00 13 15 14 05 25,

and, using the space $= 00$, A $= 01$ correspondence we agreed on, read the message

S E N D   M O N E Y.

Complexity Theory

The reader, after plowing through the calculations of the last section, may readily agree that RSA enciphering and deciphering are best left to computers. He or she may also have noticed that the difficulty of factoring our $n = 1189$, upon which the security of the method depends, is much less than that of the other calculations the method entails. Certainly if we can raise 848 to the 187 power modulo 1189, then someone else can factor 1189, and so read secret messages sent to us.

The explanation for this paradox has to do with the size of the numbers involved, and with the complexity of the calculation we must do with these numbers. Recall that in actual practice the primes p and q will be about 100 digits each, and n will be about 200 digits. It turns out that for numbers of this size the work involved in enciphering and deciphering, while not to be sneezed at, is much less than that of factoring n by any known method.

To analyze why this is so we must break down the steps involved in enciphering and deciphering into the elementary operations that make them up, for example addition, subtraction, multiplication, division, and comparison. Although the size of the numbers involved affects how hard such an operation is for a computer, to keep things simple we will assume each such operation takes the same amount of time, say one-millionth of a second.

The hardest part of deciphering seemed to be raising the numbers to the power d modulo n. The number d is the least positive solution to a congruence with modulus b, and $b = [p - 1, q - 1] = (p - 1)(q - 1)/(p - 1, q - 1) < pq/2 = n/2$, so we can assume $d < n/2$. First we converted d to binary by successively dividing it by 2. How many divisions need we perform? Clearly it is the number of digits in the binary representation of d. Say this is k. Then

$$2^{k-1} \le d, \qquad \text{or} \qquad k - 1 \le \log_2 d \le \log_2 (n/2) = (\log_2 n) - 1.$$

We see the number of divisions k is bounded by $\log_2 n$. Now $\log_2 n$ grows very slowly with n. For example when n is a 200-digit number, $\log_2 n$ is less than 700.

Note that to raise a number to the power d we first square it k times, reducing modulo n as we go. This requires fewer than 700 multiplications and 700 divisions when n has 200 digits. Now we are up to about 2100 operations. This would be big trouble for a human being, but is nothing for a computer. Finally to compute the dth power of a number we multiply the appropriate numbers we have found by successive squaring, reducing modulo n as we go. Mark down at most k more multiplications and k more divisions. Thus no more than 3500 operations are necessary.

We must not forget that in order to find d we must solve the congruence

$$ex \equiv 1 \quad (\text{mod } b).$$

We would do this by applying the Euclidean Algorithm to e and b. It is possible to show that the number of divisions needed to apply the algorithm to e and b is no more than twice the logarithm to the base 2 of the larger of these, and so less than $2 \log_2 n$, about 1400 divisions. (See the problems at the end of this section.) The coefficients that make each remainder a linear combination of e and b can be kept track of as we go along (see the last exercise in Section 1.3), making a total of, say 2800 or so operations. Thus the enciphering or deciphering of a single block of digits will involve something less than 10000 elementary operations. Our million-operation-per-second computer could knock off 100 such jobs in a second.

Perhaps one million operations per second is faster than any computer can perform with such big numbers. Still this assumption can be used for comparison purposes, for we will assume the computer being used by the people trying to break our enciphering method is equally fast.

A person trying to read the messages sent to us by factoring n will have to factor a number of 200 digits. If he tries divisors up to the square root of n, this will entail about 10^{100} divisions. Even at one million operations per second, this will take 10^{94} seconds, or about $3 \cdot 10^{86}$ years.

Of course only prime divisors of n really need be tested, but this is not as helpful as it seems. The problem is knowing which divisors are prime when the numbers involved may have as many as 100 digits. It turns out that even if the best methods presently known are used, it can be estimated that factoring a 200-digit number will take our imaginary computer about 4 billion years. By that time we won't care who reads our mail.

Why the Method Works

Why does raising the ciphertext numbers to the power d bring back the plaintext? Consider the plaintext number P. The corresponding ciphertext is $C \equiv P^e \pmod{n}$. We decipher C by computing

$$P' \equiv C^d \quad (\text{mod } n),$$

and so it suffices to show that $P' = P$. In fact since both P and P' are non-negative integers less than n, it suffices to show that

$$P' \equiv P \quad (\text{mod } n).$$

Now by the definitions of C and P' we have

$$P' \equiv C^d \equiv (P^e)^d \equiv P^{de} \pmod{n}.$$

Recall that d is the least positive solution to

$$ex \equiv 1 \pmod{b},$$

where $b = [p - 1, q - 1]$. (Since e is chosen relatively prime to $p - 1$ and $q - 1$, it is relatively prime to b and so the congruence has a solution by Theorem 4.1.)
Let $de = 1 + kb$. Now if $(p, P) = 1$, then

$$P^{p-1} \equiv 1 \pmod{p}$$

by Fermat's Theorem. Then

$$P^{kb} \equiv 1 \pmod{p}$$

since $p - 1$ divides b. Thus

$$P^{de} = P^{kb+1} \equiv P \pmod{p},$$

and this congruence is clearly true also if p divides P. In the same way

$$P^{de} \equiv P \pmod{q}.$$

Thus we have

$$P^{de} \equiv P \pmod{n}$$

by Theorem 4.2.

How Secure Are Public-Key Systems?

The security of the RSA method depends on the fact that no efficient way is known of factoring large integers. If a dramatically faster factoring method were found, then the method would become invalid. When another public-key system, called the "knapsack method," was cracked in 1982 by the Israeli mathematician Adi Shamir (the "S" of RSA), *Time* magazine ran a full-page article on the feat (see p. 144). Two years later a more complicated knapsack scheme was broken by Ernest F. Brickell of the Scandia National Laboratories; the story on p. 145 is from the Boston *Globe*. For an explanation of the knapsack method and a review of modern factoring methods the reader should consult the *Scientific American* articles by Hellman and Pomerance listed in the references.

Computers

Opening the "Trapdoor Knapsack"

An Israeli mathematician cracks a formidable code

Five years ago, computer scientists at Stanford and M.I.T. made a pair of chummy but keenly competitive $100 bets. A team at each university had devised a secret code to protect computers from electronic intruders by scrambling and unscrambling the data in a complex fashion. Each team offered cash to the first mathematician who could crack its code, figuring that the deciphering could not be done in much less than a million years. To the surprise of all concerned, however, the Stanford scheme sprang a leak this year, putting

mathematics known as complexity theory. Shamir was at M.I.T. in the late '70s as an associate professor of mathematics, and in fact helped write the M.I.T. code that competes head-on with Stanford's. Last spring, back in his spartan, second-floor office in the Weizmann Institute of Science in Rehovot, the lean, blue-jeaned mathematician settled the old wager: he found a way to unravel the original Stanford system. The code Shamir broke after four years of hard work was no Buck Rogers-Dick Tracy cipher. It was a charter member, along with

N.M. scientist cracks code

By Edward Dolnick
Globe Staff

A New Mexico computer scientist has broken a purportedly unbreakable code, foiling one of the most sophisticated schemes yet devised for keeping vital information safe from thieves and snoops.

For cracking the code, Ernest F. Brickell of the Sandia National Laboratories won a $1000 reward put up by one of the cipher's coinventors, mathematician Ralph Merkle.

"The exhilaration came when I found Merkle's check in the mail," Brickell said in a phone interview yesterday. Bets are common among the mathematicians who specialize in code-breaking, but $100 is a more typical prize than $1000, and the larger sum reflected Merkle's confidence in his cipher.

Long the province of spies and writers of thrillers, encryption—scrambling information so that outsiders cannot tap it—has lately become important in the everyday world as well.

The goal is safeguarding the growing amount of information stored in computers and protecting the flow of data transmitted by telephone or satellite. Every hour of every day, for example, an average of $13 billion is electronically transferred between banks.

Brickell, 31, worked two years on the problem known as the "knapsack" system. It is given that name, Brickell explained, because breaking the code amounts to answering a question about dividing up the contents of an enormous knapsack so they fit exactly into several smaller sacks.

Mathematically, the problem is finding which of a list of numbers add to a particular sum. For small numbers, the question can be easy: Which numbers in this list— 1, 3, 5, 8—add up to 12?

Realistic examples are harder. Breaking the knapsack cipher, Brickell said, required finding which of 100 numbers, each 60 digits long, could be added together to yield a particular 62-digit sum.

Brickell said that his code-breaking method can crack the knapsack problem in under two hours on a Cray computer, one of the most powerful machines in the world.

"You can rent Cray time at about $1000 an hour," Brickell said, "so you could break the system for $2000—and you're talking about things like Electronic Funds Transfers." But even if it took $1 million to break the code it would be worth it, so great are the potential rewards.

The code that Brickell cracked, though not in commercial use, had been widely hailed as one of the most promising yet devised. It is one of a family called "public key" ciphers, invented independently in 1976 by mathematicians at MIT and Stanford.

A less powerful version of the knapsack code was cracked in 1982 (costing Merkle $100). Another public key cipher, called RSA after the initials of its inventors, remains unbroken.

Hailed as a revolution by some historians, public key ciphers appear to have crucial advantages over conventional ciphers. Those conventional codes, though complicated in practice, are straightforward in concept.

The standard approach is akin to that in the familiar a = 1, b = 2, c = 3 codes: Messages are encrypted according to some secret "recipe" or key, and then scrambled. The message can later be decoded by going through the same procedure in reverse.

But many specialists say that such systems have crucial weaknesses. The greatest one is the difficulty of getting the secret key into all the right hands, and keeping it out of all the wrong ones.

Public key ciphers, like the knapsack system, offer a mathematical end-run around that problem.

Unlike conventional ciphers with their one secret code, each user of a public key cipher has two mathematically related keys. One key, which is used to encrypt messages, is published in a public directory; the other key, used for decryption, is kept secret.

The result is that anyone can send a secret message to a Mr. Smith, say, by looking up his key in the directory. But only Smith has the key needed to decode the message, and only he can read the secret message sent to him.

PROBLEMS FOR SECTION 4.5

Problems 1 through 10 assume the substitution method described at the beginning of this section. In problems 1 through 8 assume the key is NUMBER THEORY.

1. Encipher: SELL ON MONDAY
2. Encipher: SNODGRASS IS A DOUBLE AGENT
3. Encipher: CHERCHEZ LA FEMME
4. Encipher your name.
5. Decipher: NULPNOF RKGD UENFL
6. Decipher: PHOL OL N ROFE UGGA
7. Decipher: PHE MCGMA VOCC EWICGBE
8. Decipher: REKDNPL CNLP PHEGKED
9. Let the key be LAGRANGE.
   Encipher: EVERY INTEGER IS THE SUM OF FOUR SQUARES.
   Decipher: TCN TCNMQNJ ME WDISMK DS TQUN.
10. Let the key be EUCLID.
    Encipher: THERE ARE INFINITELY MANY PRIMES.
    Decipher: SBIQI FR MN QNYEJ QNEL SN AINKISQY.

Problems 11 through 38 refer to the RSA method. In the next four problems change each message to a sequence $P_1, \ldots, P_4$ *using the system of this section.*

11. ATTACK   12. CLAMUP   13. GET AL   14. I QUIT

In the next four problems suppose the public key is $n = 1189, e = 3$. *Convert each message to ciphertext.*

15. SPY   16. RUB   17. KGB   18. CIA

In the next four problems suppose the public key is $n = 1081, e = 3$. *Convert each message to ciphertext.*

19. WEB   20. RUM   21. FBI   22. NSA

In the next four problems compute b and d.

23. $p = 7, q = 11, e = 7$ 24. $p = 13, q = 17, e = 5$
25. $p = 19, q = 23, e = 29$ 26. $p = 31, q = 37, e = 41$

B *In the next four problems convert to binary.*

27. 53   28. 72   29. 111   30. 145

In the next four problems find the least residue of 117^e *(mod 1081).*

31. $e = 53$ 32. $e = 72$ 33. $e = 111$ 34. $e = 145$

In the next four problems decipher, given that $p = 23$, $q = 47$, and $e = 13$.

35. 1006, 961   36. 228, 714   37. 684, 950   38. 493, 229

Decipher each of the following substitution messages, and determine each key.

39. RAU IBPQR QRUL RK WBQFKH BQ GUCPJBJS RAU LPKLUP
    JCHUQ KI RABJSQ
40. D CFVA MDQRLVAPAM F SPUIY PAJFPHFEIA NPLLT WCDRC
    SCDQ JFBDK DQ SLL QJFII SL RLKSFDK

41. Show that if $0 < a < b$, and if $b = aq + r$, $0 \leq r < a$, then $r < b/2$.
42. Show that if $0 < a < b$, and if the Euclidean algorithm applied to a and b
    takes k steps (so r_k is the last nonzero remainder), then $k \leq 2 \log_2 b$.

Carl Friedrich Gauss

1777–1855

If a poll were taken among mathematicians as to who was the all-time greatest practitioner of the subject, Gauss would probably get more votes than anyone else. The son of a German bricklayer, Gauss conceived most of his discoveries before the age of 20, but spent the rest of his life polishing and refining them.

His motto, "few, but ripe," reflected his habit of retaining, even concealing, his results until he thought them sufficiently perfected (or until he felt the world was ready for them). For example Gauss laid the foundations of non-Euclidean geometry, but hid his researches until others published the idea.

Gauss considered number theory to be the most interesting part of mathematics, but spent much of his energies on more applied subjects. He invented the method of least squares independently from Legendre, and, in his position as director of the astronomical observatory at Göttingen, calculated the orbits of heavenly bodies. He also worked in electricity and magnetism, inventing with

CARL FRIEDRICH GAUSS Wilhelm Weber a working telegraph in 1833.

Early in his life Gauss had to choose between a career in mathematics or the study of languages. His discovery, at the age of 19, of the constructability of the regular 17-gon, led him to choose the former. This was the first advance along this line since the time of Euclid, more than 2000 years earlier.

Gauss was also the first to put complex numbers on a logical footing, and he introduced to analysis what today is considered mathematical rigor.

Gauss's life was long and generally quiet and happy. He disliked travel, but retained his interest in literature and world affairs, reading extensively in English and Russian. His accomplishments, both in and out of mathematics, were so extensive that many must be omitted in a short account such as this one.

Chapter 5

Congruences of Higher Degree

Just as in algebra one goes from the study of linear equations to those involving higher powers of the unknown, so also can we proceed from the study of linear congruences to those of higher degree. Complications arise, however, as we will see.

5.1 Polynomial Congruences

We will consider polynomial congruences in a single unknown, for example

$$5x^4 + 17x^3 - 3x + 2 \equiv 2x^3 - 7 \pmod{18}.$$

In fact we will only consider congruences that can be put in the form

$$F(x) \equiv 0 \pmod{m}, \tag{$*$}$$

where $F(x)$ is a polynomial in x with integer coefficients. Our example could be put in this form by shifting everything to the left of the congruence sign, so that

$$F(x) = 5x^4 + 15x^3 - 3x + 9.$$

By Theorem 1.8 if $a \equiv a' \pmod{m}$, then $ax^k \equiv a'x^k \pmod{m}$ for all integers x. Thus coefficients of $F(x)$ may be replaced by congruent coefficients without changing the set of solutions of $(*)$. For example,

$$5x^4 + 15x^3 - 3x + 9 \equiv 0 \pmod{9}$$

149

is equivalent to

$$5x^4 + 6x^3 - 3x \equiv 0 \pmod 9.$$

Any term of $F(x)$ having a coefficient divisible by the modulus may be dropped completely, since it is congruent to 0 no matter what x is. These considerations motivate the following definition.

DEFINITION   degree of a congruence
If $F(x)$ is a polynomial in x with integral coefficients, by the **degree** of the congruence

$$F(x) \equiv 0 \pmod m$$

we mean the highest power of x in $F(x)$ whose coefficient is not divisible by m.

Examples   Let $F(x) = 6x^3 + x^2 + 8$. Then the degree of

$$F(x) \equiv 0 \pmod m$$

is 3 if $m = 5$, but if $m = 3$ then the degree is only 2.
For $m = 5$ we can find a complete solution by testing the values $x = 0, 1, 2, 3$, and 4 as follows:

x	0	1	2	3	4
$F(x)$	8	15	60	179	408

Thus a complete solution is $x = 1, 2$. (Actually it would be easier to test the values $x = -2, -1, 0, 1, 2$.) Another complete solution is $x = 42, 16$ (since $42 \equiv 2$ and $16 \equiv 1 \pmod 5$.)

Reducing a Congruence to Prime Power Moduli

Consider the congruence

$$F(x) = 6x^3 + x^2 + 8 \equiv 0 \pmod{20}.$$

Solving this by trial-and-error would involve evaluating $F(x)$ for 20 integers, a tedious task. There is a way to reduce the work, however. Recall that Theorem 4.2 says that if $(a, b) = 1$, then

$$z \equiv z' \pmod{ab} \quad \text{if and only if} \quad z \equiv z' \pmod a \quad \text{and} \quad z \equiv z' \pmod b.$$

Applying this theorem with $z = F(x)$, $z' = 0$, $a = 5$, and $b = 4$, we see that

$$F(x) \equiv 0 \pmod{20}$$

is equivalent to

$$F(x) \equiv 0 \pmod 5 \quad \text{and} \quad F(x) \equiv 0 \pmod 4.$$

We solved the congruence with modulus 5 in the last example, finding that $x = 1, 2$ was a complete solution. From the table of values of $F(x)$ above we see that $x = 0, 2$ is a complete solution for the modulus 4. Then x is a solution to the original congruence if and only if

$$x \equiv 1 \text{ or } 2 \pmod 5 \quad \text{and} \quad x \equiv 0 \text{ or } 2 \pmod 4. \qquad (**)$$

Solving these simultaneous congruences can be done by the Chinese Remainder Theorem, discussed in Sections 4.2 and 4.3. Each pair of solutions modulo 4 and 5 generates a unique solution modulo 20, so we get four (2 times 2) solutions in all. The reader should review the Chinese Remainder Theorem and confirm that x satisfies $(**)$ if and only if $x \equiv 2, 6, 12,$ or $16 \pmod{20}$. Thus this is a complete solution to the original congruence.

Notice that in order to apply Theorem 4.2 and the Chinese Remainder Theorem the smaller moduli employed must be relatively prime to each other. Thus factoring 20 into 4 times 5 worked, but 2 times 10 would not have, since these are not relatively prime.

This method may be used to simplify the solution of

$$F(x) \equiv 0 \pmod m$$

whenever m can be factored into relatively prime smaller factors, that is, whenever more than one prime divides m. More than two factors may be used. For example $F(x) \equiv 0 \pmod{360}$ may be replaced by

$$F(x) \equiv 0 \pmod 9 \quad \text{and} \quad F(x) \equiv 0 \pmod{40}.$$

But the latter congruence is equivalent to

$$F(x) \equiv 0 \pmod 8 \quad \text{and} \quad F(x) \equiv 0 \pmod 5.$$

The smallest pieces the modulus can be broken into are the prime powers in its factorization. This method is summarized in the following theorem. The details of the proof (a version of Theorem 4.2 for more than two moduli is needed, for example) are left for the exercises.

THEOREM 5.1 Consider the congruence

$$F(x) \equiv 0 \pmod m,$$

where $F(x)$ is a polynomial with integer coefficients. Let

$$m = p_1^{n_1} \cdots p_r^{n_r},$$

where the p's are distinct primes. Suppose a complete solution to

$$F(x) \equiv 0 \pmod{p_i^{n_i}}$$

has k_i elements, $i = 1, 2, \ldots, r$. Then a complete solution to the original congruence has $k_1 k_2 \cdots k_r$ elements. In fact, if for each i, $i = 1, \ldots, r$, x_i is a solution to

$$F(x) \equiv 0 \pmod{p_i^{n_i}},$$

then any x such that

$$x \equiv x_i \pmod{p_i^{n_i}}, \quad i = 1, \ldots, r$$

satisfies the original congruence, and a complete solution to it may be constructed this way by allowing the x_i to run through complete solutions to the prime power moduli.

Example   Solve $3x^2 - 20x + 25 \equiv 0 \pmod{84}$.

The prime power factors of 84 are 4, 3, and 7. By testing values in complete residue systems we find that a complete solution to

$$3x^2 + 1 \equiv 0 \pmod 4$$

is $x = 1, 3$; a complete solution to

$$x + 1 \equiv 0 \pmod 3$$

is $x = -1$; and a complete solution to

$$3x^2 + x + 4 \equiv 0 \pmod 7$$

is $x = -3, -2$. (Note that the congruences have been simplified by reducing coefficients, depending on the modulus.)

Now we use the Chinese Remainder Theorem to solve the simultaneous congruences

$$x \equiv 1 \text{ or } 3 \pmod 4$$

$$x \equiv -1 \pmod 3$$

$$x \equiv -3 \text{ or } -2 \pmod 7,$$

which involves solving

$$21x_1 \equiv 1 \quad (\text{mod } 4),$$

$$28x_2 \equiv 1 \quad (\text{mod } 3), \qquad \text{and}$$

$$12x_3 \equiv 1 \quad (\text{mod } 7).$$

Solutions are $x_1 = 1$, $x_2 = 1$, and $x_3 = 3$. Then the simultaneous solutions are

$$x = 21(1)(1 \text{ or } 3) + 28(1)(-1) + 12(3)(-3 \text{ or } -2)$$
$$= -115, -79, -37, -73.$$

This is a complete solution to the original congruence. The least complete solution is $x = 5, 11, 47, 53$.

Example Solve $x^2 + 1 \equiv 0 \quad (\text{mod } 35)$.
 By trial and error we find that of the two congruences

$$x^2 + 1 \equiv 0 \quad (\text{mod } 5) \qquad \text{and} \qquad x^2 + 1 \equiv 0 \quad (\text{mod } 7),$$

$x = -2, 2$ is a complete solution to the first, but the second congruence has no solutions. Thus the original congruence has no solutions either. This illustrates that in Theorem 5.1 some of the numbers k_i may be zero.

Congruences with Modulus p^2

The above method does not help with the congruence

$$2x^3 - 3x + 6 \equiv 0 \quad (\text{mod } 25), \tag{1}$$

since only one prime divides the modulus. It is possible to avoid testing 25 values of x in it, however. The method we will illustrate depends on the fact that by part (f) of Theorem 1.8 any solution to (1) also satisfies

$$2x^3 - 3x + 6 \equiv 0 \quad (\text{mod } 5). \tag{2}$$

By testing $x = -2, -1, 0, 1$, and 2 in (2) we see it has a complete solution $x = 1$.
 Thus instead of trying all of $0, 1, \ldots, 24$ in (1) we need only try values of $x \equiv 1 \ (\text{mod } 5)$, namely $x = 1, 6, 11, 16$, and 21.
 Even these computations may be avoided. We want to test solutions of the form $x = 1 + 5y$ in (1), where $y = 0, 1, 2, 3$, or 4. Substituting this expression in (1) gives

$$2(1 + 5y)^3 - 3(1 + 5y) + 6 \equiv 0 \quad (\text{mod } 25).$$

Simplifying this algebraically and dropping those terms with coefficients divisible by 25 yields

$$15y + 5 \equiv 0 \quad (\text{mod } 25).$$

By Theorem 1.10 (with $a = 5$ and $b = 25$) this is equivalent to

$$3y + 1 \equiv 0 \quad (\text{mod } 5). \tag{3}$$

Since we are interested in y ranging from 0 to 4, we want a complete solution. One is easily seen to be $y = 3$. We see that a complete solution to (1) is $x = 1 + 5(3) = 16$.

The method of this example works in general. That is, given a congruence

$$F(x) \equiv 0 \quad (\text{mod } p^2), \tag{1'}$$

where p is a prime, we first solve

$$F(x) \equiv 0 \quad (\text{mode } p). \cdot \tag{2'}$$

Now if x' is any solution to (2') we look for solutions of (1') of the form $x = x' + py$, where y is in some complete residue system (mod p). Substituting this expression in (1') leads to a *linear* congruence

$$ay + b \equiv 0 \quad (\text{mod } p) \tag{3'}$$

in y, to which a complete solution is desired. The reason that (3') turns out to be linear is that when $x' + py$ is raised to a power, most of the terms involving y drop out because they have a coefficient divisible by p^2.

Note that if (2') has more than one solution x' the succeeding process must be applied to each of them. This technique will be generalized and investigated in detail in the next section.

PROBLEMS FOR SECTION 5.1

 In the first ten problems find the least complete solution by testing.

1. $x^3 - x^2 + 3x + 1 \equiv 0 \quad (\text{mod } 7)$ 2. $x^3 - x^2 + 3x + 1 \equiv 0 \quad (\text{mod } 5)$
3. $x^2 - 3x + 1001 \equiv 0 \quad (\text{mod } 3)$ 4. $x^2 - 3x + 1001 \equiv 0 \quad (\text{mod } 5)$
5. $x^2 - 3x + 1001 \equiv 0 \quad (\text{mod } 7)$ 6. $x^4 - x^3 + 5x + 1 \equiv 0 \quad (\text{mod } 3)$
7. $x^4 - x^3 + 5x + 1 \equiv 0 \quad (\text{mod } 11)$ 8. $x^4 - x^3 + 5x + 1 \equiv 0 \quad (\text{mod } 4)$
9. $x^7 - x \equiv 0 \quad (\text{mod } 7)$ 10. $x^7 - x \equiv 0 \quad (\text{mod } 5)$

In the next four problems find all solutions x with $|x| \leq 10$.

11. $x^3 + x + 3 \equiv 0 \quad (\text{mod } 7)$ 12. $x^3 + x + 3 \equiv 0 \quad (\text{mod } 13)$
13. $x^3 + x + 3 \equiv 0 \quad (\text{mod } 19)$ 14. $x^3 + x + 3 \equiv 0 \quad (\text{mod } 3)$

In the next ten problems break the modulus into prime powers to find the least complete solution.

15. $x^3 - x^2 + 3x + 1 \equiv 0 \pmod{35}$

16. $x^2 - 3x + 1001 \equiv 0 \pmod{15}$

17. $x^2 - 3x + 1001 \equiv 0 \pmod{10}$

18. $x^2 - 3x + 1001 \equiv 0 \pmod{21}$

19. $x^4 - x^3 + 5x + 1 \equiv 0 \pmod{33}$

20. $x^4 - x^3 + 5x + 1 \equiv 0 \pmod{66}$

21. $x^2 + x + 1 \equiv 0 \pmod{21}$

22. $x^2 + x + 1 \equiv 0 \pmod{91}$

23. $x^2 + x + 1 \equiv 0 \pmod{273}$

24. $x^2 + x + 1 \equiv 0 \pmod{195}$

B | *In the next four problems a congruence is given with modulus p^2, p prime. A solution x' to the corresponding congruence with modulus p is also given. Substitute $x = x' + py$ into the congruence and find a linear congruence with modulus p that y must satisfy as in the example at the end of this section. Find any corresponding solutions x, $0 \le x < p^2$.*

25. $x^2 + x + 1 \equiv 0 \pmod{49}$, $x' = 2$

26. $x^2 + x + 1 \equiv 0 \pmod{49}$, $x' = 4$

27. $x^3 + 8 \equiv 0 \pmod{25}$, $x' = 3$

28. $x^3 + 8 \equiv 0 \pmod{9}$, $x' = 1$

In the next four problems use the method at the end of this section to find the least complete solution.

29. $x^2 - x + 2 \equiv 0 \pmod{25}$

30. $x^2 - x + 2 \equiv 0 \pmod{49}$

31. $x^2 + x + 2 \equiv 0 \pmod{121}$

32. $x^3 + 8 \equiv 0 \pmod{49}$

C | 33. Let k and k' be the number of elements in complete solutions to $F(x) \equiv 0 \pmod 3$ and $F(x) \equiv 0 \pmod 9$, respectively. Construct integral polynomials $F(x)$ such that
   (a) $k = 3$, $k' = 0$ (b) $k = 1$, $k' = 3$
   (c) $k = 1$, $k' = 1$ (d) $k = 2$, $k' = 4$

34. Use induction on n to prove that if $m = m_1 \cdots m_n$ and the m_i are relatively prime in pairs, then $z \equiv z' \pmod m$ if and only if $z \equiv z' \pmod{m_i}$ for $i = 1, \ldots, n$.

35. Prove Theorem 5.1.

5.2   Congruences with Prime Power Moduli

Generalizing the Method of the Last Section

In the last section we saw how solving any polynomial congruence can be reduced, with the help of the Chinese Remainder Theorem, to solving congruences with moduli powers of primes; and at the end of the section a method of solving congruences with modulus p^2 was illustrated. We will generalize the latter method to solve congruences with modulus p^n, where p is prime.

The general plan, given a congruence

$$F(x) \equiv 0 \pmod{p^n},$$

will be to find first a complete solution to

$$F(x) \equiv 0 \pmod{p},$$

then use this to solve

$$F(x) \equiv 0 \pmod{p^2},$$

and continue this way until we have a complete solution to the original congruence. What we need is a method for going from modulus p^k to modulus p^{k+1}.
   Consider the congruences

$$F(x) \equiv 0 \pmod{p^k}, \tag{$*$}$$

and

$$F(x) \equiv 0 \pmod{p^{k+1}}. \tag{$**$}$$

By the last part of Theorem 1.8 any solution of $(**)$ is also a solution of $(*)$, so if $x_k, x_k', x_k'', \ldots$ is a complete solution of $(*)$, then any solution of $(**)$ will be congruent to one of these numbers $\pmod{p^k}$.
   Thus we can restrict our attention to solutions of $(**)$ of the form

$$x_k + p^k y, x_k' + p^k y, \ldots,$$

where y is an integer. Furthermore, we can assume that y is in some complete residue system $\pmod{p}$, since if $y \equiv y' \pmod{p}$, then it can be checked that

$$x_k + p^k y \equiv x_k + p^k y' \pmod{p^{k+1}},$$

and so these numbers could not both appear in a complete solution to $(**)$.
   Let us consider a particular solution x_k of $(*)$ and see what solutions of the form $x_k + p^k y$ of $(**)$ it might generate. We wish to find out which y satisfy

$$F(x_k + p^k y) \equiv 0 \pmod{p^{k+1}}. \tag{$***$}$$

Let us assume

$$F(x) = a_r x^r + a_{r-1} x^{r-1} + \cdots + a_0.$$

Then a typical term on the left side of (∗∗∗) will be

$$a_j(x_k + p^k y)^j.$$

If this is multiplied out, an unpleasant expression of $j + 1$ terms will result; but fortunately it turns out that all but the first two will be divisible by p^{k+1}, and so may be dropped from the congruence (∗∗∗).

Since the complexity of the expression involved may obscure what is going on, let us simplify the notation. We claim that

$$(u + mv)^j = u^j + ju^{j-1}mv + (\text{a multiple of } m^2)$$

for any integers u, m, and v, and any positive integer j. Although the Binomial Theorem could be invoked to prove this, all that is really needed is to notice that the expression on the left is $u + mv$ multiplied by itself j times, that is

$$\underbrace{(u + mv)(u + mv) \cdots (u + mv)}_{j \text{ factors}}.$$

Forming a term by choosing all u's gives u^j. Choosing one factor mv and the rest u's can be done in j ways, and so gives a term $ju^{j-1}mv$. All other terms contain the factor mv at least twice, and so are multiples of m^2.

In particular, if we take $u = x_k$, $m = p^k$, and $v = y$, we see that

$$a_j(x_j + p^k y)^j = a_j(x_k^j + jx_k^{j-1}p^k y + \text{a multiple of } p^{2k}).$$

Since it is easy to see that $2k \geq k + 1$ for k any positive integer, this means that in the congruence (∗∗∗) the term in question may be replaced by

$$a_j x_k^j + a_j j x_k^{j-1} p^k y.$$

By collecting the terms involving y we see that (∗∗∗) may be written in the form

$$F(x_k) + F'(x_k)p^k y \equiv 0 \quad (\text{mod } p^{k+1}),$$

where $F'(x)$ is the polynomial

$$a_r r x^{r-1} + a_{r-1}(r-1)x^{r-2} + \cdots + a_1.$$

Calculus students should recognize that this is just the derivative of $F(x)$. No calculus is really needed here, however, since it is easy enough to define $F'(x)$ in a formal way for any polynomial $F(x)$.

DEFINITION   derivative of a polynomial

Given a polynomial $F(x) = a_r x^r + \cdots + a_0$, we define its (formal) **derivative** to be the polynomial

$$F'(x) = a_r r x^{r-1} + a_{r-1}(r-1)x^{r-2} + \cdots + a_1.$$

For example the derivative of

$$3x^5 - 4x^3 + 2x^2 + 9x + 14$$

is

$$15x^4 - 12x^2 + 4x + 9.$$

A Linear Congruence for y

We saw above that if x_k is a solution to (∗), then $x_k + p^k y$ is a solution to (∗∗) exactly when

$$F(x_k) + F'(x_k)p^k y \equiv 0 \quad (\text{mod } p^{k+1}).$$

By the Cancellation Theorem this congruence is equivalent to

$$\frac{F(x_k)}{p^k} + F'(x_k)y \equiv 0 \quad (\text{mod } p),$$

or

$$F'(x_k)y \equiv \frac{-F(x_k)}{p^k} \quad (\text{mod } p).$$

(Note that the term on the right is an integer by the assumption that x_k satisfies (∗).) Thus finding y involves only solving a linear congruence with modulus p.

Before summarizing this method in a theorem we will illustrate it.

Example   Solve $x^2 + x + 3 \equiv 0 \pmod{27}$.

We start by solving the congruence

$$x^2 + x + 3 \equiv 0 \quad (\text{mod } 3)$$

by testing values of x in a complete residue system (mod 3), and find a complete solution is $x_1 = -1, 0$.

Solutions to the congruence with modulus 9 will be of the form $x_1 + 3y$, where y satisfies

$$F'(x_1) \equiv \frac{-F(x_1)}{3} \quad (\text{mod } 3).$$

Note that $F'(x) = 2x + 1$. For $x_1 = -1$ this congruence is

$$(-1)y \equiv \frac{-3}{3} \quad \text{(mod 3)}.$$

This has the complete solution $y = 1$, giving $x_2 = -1 + 3(1) = 2$.
If we take $x_1 = 0$ then the congruence is

$$(1)y \equiv \frac{-3}{3} \quad \text{(mod 3)},$$

which has the complete solution $y = -1$, giving $x_2 = 0 + 3(-1) = -3$.
  Now we look for solutions to the original congruence with modulus 27 of the form $x_3 = x_2 + 9y$. Here y must satisfy

$$F'(x_2) \equiv \frac{-F(x_2)}{9} \quad \text{(mod 3)}.$$

For $x_2 = 2$ this congruence is

$$5y \equiv \frac{-9}{9} \quad \text{(mod 3)},$$

which has the complete solution $y = 1$; and so $x_3 = 2 + 9(1) = 11$.
  Using $x_2 = -3$ gives the congruence

$$-5y \equiv \frac{-9}{9} \quad \text{(mod 3)},$$

which has the complete solution $y = -1$. Thus here $x_3 = -3 + 9(-1) = -12$. A complete solution to the original congruence is $x = 11, -12$. The least complete solution is $x = 11, 15$.

THEOREM 5.2 Suppose $x_k, x_k', x_k'', \ldots$ is a complete solution to $F(x) \equiv 0 \pmod{p^k}$, where $F(x)$ is an integral polynomial and p is a prime. Then all solutions to $F(x) \equiv 0 \pmod{p^{k+1}}$ in some complete residue system modulo p^{k+1} congruent to $x_k \pmod{p^k}$ are given by $x_{k+1} = x_k + p^k y$, where y runs through any complete solution to

$$F'(x_k)y \equiv \frac{-F(x_k)}{p^k} \quad \text{(mod } p\text{)}.$$

Applying this also to $x_k', x_k'', \ldots$ in turn yields a complete solution to $F(x) \equiv 0 \pmod{p^{k+1}}$.

Notice that according to Theorem 4.1 the congruence $ax \equiv c \pmod{b}$ has a solution if and only if (a, b) divides c, and if solutions exist, then a complete solution has (a, b) elements. In the case of the congruence defining y we have the modulus b is p, and so (a, b) equals 1 or p. Thus each solution x_k generates 0, 1, or p solutions x_{k+1} at the next level.

Example Solve $x^3 + x^2 + 23 \equiv 0 \pmod{125}$.

We start with the congruence $F(x) \equiv 0 \pmod 5$, where $F(x) = x^3 + x^2 + 23$. The following table shows that a complete solution is $x_1 = 1, 2$.

x_1	-2	-1	0	1	2
$F(x_1)$	19	23	23	25	35

Notice that $F'(x) = 3x^2 + 2x$. We will find solutions x_2 to $F(x) \equiv 0 \pmod{25}$ of the form $x_1 + 5y$, where y satisfies

$$F'(x_1) \equiv \frac{-F(x_1)}{5} \pmod 5.$$

For $x_1 = 1$ this congruence is

$$5y \equiv \frac{-25}{5} \pmod 5,$$

and a complete solution is $y = -2, -1, 0, 1, 2$. Thus $x_2 = 1 + 5y = -9, -4, 1, 6, 11$.

For $x_1 = 2$ the congruence for y becomes

$$16y \equiv \frac{-35}{5} \pmod 5,$$

and a complete solution is $y = -2$, yielding $x_2 = 2 + 5y = -8$.

It is useful to make a table of F and F' for the various values of x_2.

x_2	-9	-4	1	6	11	-8
$F(x_2)$	-625	-25	25	275	1475	-425
$F'(x_2)$			$\equiv 5 \pmod 5$			$\equiv 16 \pmod 5$

Notice that in the congruence

$$F'(x_k)y \equiv \frac{-F(x_k)}{p^k} \pmod p$$

all that matters about $F'(x_k)$ is what it is modulo p. Since all solutions x_k arising from a given solution x_1 at the first level will be congruent (mod p), F' need only be computed once for each of these. For example all the solutions $x_2 = -9$, $-4, 1, 6, 11$ arising from $x_1 = 1$ will have $F'(x_2) \equiv 5$ (mod 5).

We must find new values of y satisfying

$$F'(x_2)y \equiv \frac{-F(x_2)}{25} \quad (\text{mod } 5).$$

For $x_2 = -9, -4, 1, 6, 11$ we get the congruences

$$5y \equiv 25, 1, -1, -11, -59 \quad (\text{mod } 5).$$

Only the first of these is solvable, with the complete solution $y = -2, -1, 0$, 1, 2, which yields $x_3 = -9 + 25y = -59, -34, -9, 16, 41$.

Taking $x_2 = -8$ leads to the congruence

$$16y \equiv 17 \quad (\text{mod } 5)$$

with complete solution $y = 2$, which gives $x_3 = -8 + 25y = 42$.

We see that a complete solution to the original congruence is $x = -59, -34$, $-9, 16, 41, 42$. The least complete solution is $x = 16, 41, 42, 66, 91, 116$.

PROBLEMS FOR SECTION 5.2

A *In the first six problems a polynomial $F(x)$ is given. Find the derivative $F'(x)$.*

1. $5x^2 + 3x - 7$ 2. $-3x^4 - 2x^2 + 4x + 2$
3. $7x^{10} - 2x^7 + 3x$ 4. $9x^8 + 2x^5 - x + 13$
5. $(x + 1)^3$ 6. $(x + 2)^4$

In the next twelve problems find the least complete solution.

7. $x^2 - x + 7 \equiv 0$ (mod 9)   8. $x^2 + 3x - 50 \equiv 0$ (mod 25)
9. $x^2 - x + 7 \equiv 0$ (mod 27)   10. $x^2 + 3x - 50 \equiv 0$ (mod 125)
11. $x^2 - 44 \equiv 0$ (mod 27)   12. $x^2 - 44 \equiv 0$ (mod 81)
13. $x^3 + x^2 + 10 \equiv 0$ (mod 32)   14. $x^3 + x^2 + 1 \equiv 0$ (mod 27)
15. $x^3 + x^2 + 1 \equiv 0$ (mod 81)   16. $x^4 + x + 23 \equiv 0$ (mod 25)
17. $x^4 + x + 23 \equiv 0$ (mod 125)   18. $x^3 + 13 \equiv 0$ (mod 49)

B 19. Show that if $y \equiv y'$ (mod p), then $x_k + p^k y \equiv x_k + p^k y'$ (mod p^{k+1}), as is claimed in the proof of Theorem 5.2.
20. Prove by induction on n that if m and n are positive integers, then $(u + mv)^n \equiv u^n + nu^{n-1}mv$ (mod m^2).

In the next six problems find the least complete solution.

21. $x^2 - x - 12 \equiv 0 \pmod{100}$ ·
22. $10x^2 - x^3 \equiv 9 \pmod{405}$
23. $x^2 - x - 12 \equiv 0 \pmod{1000}$
24. $x^2 + 3x - 50 \equiv 0 \pmod{275}$
25. $x^2 - x + 7 \equiv 0 \pmod{63}$
26. $x^2 - x + 7 \equiv 0 \pmod{117}$

27. Find all x, $0 \le x < 72$, such that $x^2 + 3x + 6 \equiv 0 \pmod 8$ and $x^3 + x^2 + 3 \equiv 0 \pmod 9$.

28. Suppose F is an integral polynomial, p is a prime, and $F(x_0) \equiv 0 \pmod{p^n}$ for all positive integers n. Prove that $F(x_0) = 0$.

29. Suppose F is an integral polynomial, p is a prime, and S is a finite set such that given $n > 0$ there exists x_n in S such that $F(x_n) \equiv 0 \pmod{p^n}$. Show there exists x_0 in S such that $F(x_0) = 0$.

C

30. Prove the **Binomial Theorem** by induction on n:

$$(a + b)^n = a^n + C(n, 1)a^{n-1}b + C(n, 2)a^{n-2}b^2 + \cdots + C(n, n - 1)ab^{n-1} + b^n$$

for all positive integers n, where $C(n, r) = n!/r!(n - r)!$.

31. Since $(a + b)^n$ is clearly an integral polynomial, the Binomial Theorem shows that $n!/r!(n - r)!$ is an integer for $1 < r < n$. A direct proof is not so easy to come by. Criticize the following argument: The r consecutive integers $n, n - 1, \ldots, n - r + 1$ contain a complete residue system modulo k for $1 \le k \le r$. Thus one of them is divisible by k. Since this works for $k = 1, 2, \ldots, r$ we conclude that $r!$ divides $n(n - 1) \cdots (n - r + 1) = n!/(n - r)!$.

32. Show that if a and b are positive integers and $b = aq + r$, $0 \le r < a$, then exactly q of the integers $1, 2, \ldots, b$ are divisible by a.

33. Suppose p is a prime, b is a positive integer, and $b = q_i p^i + r_i$, $0 \le r_i < p^i$ for $i = 1, 2, \ldots$. Let p^t divide $b!$ exactly. Prove that $t = q_1 + q_2 + \cdots$. (See Section 2.3 for the definition of dividing exactly.)

34. Suppose a, b, and b' are positive integers, with $b + b' = b''$. Let $b = aq + r$, $b' = aq' + r'$, and $b'' = aq'' + r''$, with $0 \le r$, r', $r'' < a$. Show that $q + q' \le q''$.

35. Suppose p is prime and b and b' are positive integers, with $b + b' = b''$. Suppose p^t and $p^{t'}$ exactly divide $b!$ and $b'!$, respectively. Show that $p^{t+t'}$ divides $b''!$.

36. Show that if n is a positive integer and $0 \le r \le n$, then $r!(n - r)!$ divides $n!$.

37. Show that $3^{48} \| 100!$.

38. Solve $2^x \| 100!$.

39. Solve $7^y \| 462!$.

40. Solve $6^z | 45!$, $6^{z+1} \nmid 45!$.

41. Prove or disprove: If p and q are primes, $p < q$, and if $q^k | n!$, then $p^k | n!$.

42. Prove that if n and r are nonnegative integers, then $r!$ divides $(n + 1)(n + 2) \cdots (n + r)$.

In the next three problems assume that F is an integral polynomial and p is a prime.

43. Prove or disprove: If $F(x) \equiv 0 \pmod{p^n}$ is solvable for all positive integers n, then there exists an x_0 such that $F(x_0) = 0$.

44. Suppose $p \mid F(x_0)$ but $p \nmid F'(x_0)$. Show that $F(x) \equiv 0 \pmod{p^n}$ has a solution for all positive integers n.

45. Let h be the number of elements in a complete solution to $F(x) \equiv 0 \pmod{p^2}$ and suppose $p \nmid h$. Show that $F(x) \equiv 0 \pmod{p^n}$ has a solution for all positive integers n.

46. For what integers h does there exist an integral polynomial F such that a complete solution to $F(x) \equiv 0 \pmod 9$ has exactly h elements?

5.3 Quadratic Residues

Congruences of Degree Two

In the previous two sections we have seen how to reduce the solution of any polynomial congruence to that of congruences with prime moduli. For such congruences, however, our only technique is to test the elements of a complete residue system.

A reasonable way to start the analysis of such congruences would be to restrict their degree. Since linear congruences are completely covered in Section 4.1, we will consider congruences of degree two, say

$$ax^2 + bx + c \equiv 0 \pmod p, \qquad (*)$$

where p is prime.

The first thing that comes to mind on looking at $(*)$ is the quadratic formula. This says that the solutions of the *equation $ax^2 + bx + c = 0$* are given by

$$x = \frac{-b \pm \sqrt{b^2 - 4ac}}{2a}.$$

If we are interested only in real solutions there will be 0, 1, or 2 of them according as $b^2 - 4ac$ is negative, zero, or positive.

In the congruence $(*)$ we are looking for integral solutions x. Of course if $b^2 - 4ac$ is not a perfect square, then the radical will not be an integer (it may not even be real). Perhaps it would be sufficient for $b^2 - 4ac$ to be *congruent* to a square, say y^2, $\pmod p$. Then we might take

$$x = \frac{-b + y}{2a}.$$

A problem still remains, namely the division by $2a$, which may not produce an integer. Why not just require x to be a solution to

$$2ax \equiv -b + y \pmod{p}?$$

By Theorem 4.1 this will exist whenever $(2a, p) = 1$. But we can assume that p does not divide a, since otherwise (∗) is not really a congruence of degree 2. Also if p divides 2, then $p = 2$. We should be so lucky; solving congruences with modulus 2 is no great problem.

Our musings have led us to the notion that if $y^2 \equiv b^2 - 4ac \pmod{p}$, and if x is a solution to $2ax \equiv -b + y \pmod{p}$, then x is a solution to the congruence (∗). We might even dream that every solution to (∗) could be generated this way.

The strange thing is that it works! We have just seen an example of a type of reasoning, compounded of intuition, guesswork, and wishful thinking, that pervades mathematical creation yet almost never gets into print. Although sometimes completely fruitless, such uncritical thinking often carries one close enough to the truth to direct the application of more rigorous arguments, just as an illegal wiretap, while not admissible in court, may lead the police to construct a legitimate case.

THEOREM   The solutions of the congruence

$$ax^2 + bx + c \equiv 0 \pmod{p},$$

where p is an odd prime not dividing a, are given by the solutions to $2ax \equiv -b + y \pmod{p}$, where y runs through a complete solution to $y^2 \equiv b^2 - 4ac$. A complete solution has 1 element if p divides $b^2 - 4ac$, otherwise 0 or 2 elements.

Proof   To prove the first sentence we must show two things:

1. If $y^2 \equiv b^2 - 4ac$ and $2ax \equiv -b + y \pmod{p}$, then x is a solution to the original congruence.
2. Every solution to the original congruence is generated this way.

As with the proof of the quadratic formula proving (1) merely involves substitution into the original congruence, and will be left for the exercises at the end of this section.

To prove (2) we also copy the corresponding part of the proof of the quadratic formula, which involves completing the square. Let us assume x satisfies the original congruence. We multiply by $4a$ to be sure that everything remains integral when we complete the square, getting

$$4a^2x^2 + 4abx + 4ac \equiv 0 \pmod{p},$$

or

$$(2ax + b)^2 - b^2 + 4ac \equiv 0 \quad (\text{mod } p).$$

If we now define y to be $2ax + b$ then we have from the last congruence that $y^2 \equiv b^2 - 4ac \pmod{p}$, and also that $2ax \equiv -b + y \pmod{p}$ from the definition of y.

Notice that by Theorem 4.3 if y generates x and y' generates x' by this method, then $x \equiv x' \pmod{p}$ if and only if $y \equiv y' \pmod{p}$. Thus to count the elements in a complete solution to the original congruence it suffices to count a complete solution to $y^2 \equiv b^2 - 4ac \pmod{p}$. Now if y is a solution and y' is any other, then $y^2 \equiv y'^2 \pmod{p}$, so p divides $y^2 - y'^2 = (y - y')(y + y')$. By Theorem 2.1 this implies $y' \equiv y$ or $-y \pmod{p}$. There are two solutions unless $y \equiv -y \pmod{p}$, which says p divides $2y$. Since p is odd the latter happens only when p divides y (and so y^2), in which case there is only one solution. But $y^2 \equiv b^2 - 4ac \pmod{p}$, which proves the last sentence of the theorem.

Example Solve $x^2 + 6x + 1 \equiv 0 \pmod{31}$.
Here $a = 1$, $b = 6$, and $c = 1$. The obvious solutions to

$$y^2 \equiv b^2 - 4ac = 32 \equiv 1 \quad (\text{mod } 31)$$

are $y = \pm 1$. Then we must solve

$$2ax \equiv -b + y \quad \text{or} \quad 2x \equiv -6 \pm 1 = -5 \text{ or } -7 \quad (\text{mod } 31).$$

Solutions are $x = 13$, 12; and this is a complete solution to the original congruence.

Quadratic Residues

Clearly it is of interest to know for which numbers m the congruence $y^2 \equiv m \pmod{p}$ is solvable. If p divides m then, as we have seen, $y = 0$ is a complete solution, but if $p \nmid m$ whether a solution exists may not be obvious.

DEFINITION quadratic residue and quadratic nonresidue
Let p be prime and suppose $p \nmid a$. We call a a **quadratic residue modulo** p in case there exists an integer y such that $y^2 \equiv a \pmod{p}$. If no such y exists then we call a a **quadratic nonresidue modulo** p.

Examples The integers 1, 4, and 12 are quadratic residues (mod 13) (note that $5^2 \equiv 12 \pmod{13}$), while 2 is a quadratic nonresidue (mod 5) since $y^2 \equiv 2 \pmod{5}$ has no solution.

Since if $a \equiv a'$ (mod p), then the congruences

$$y^2 \equiv a \quad (\text{mod } p) \qquad \text{and} \qquad y^2 \equiv a' \quad (\text{mod } p)$$

are equivalent, the quadratic residues make up whole congruence classes. The argument at the end of the proof of the previous theorem shows that if p is an odd prime, if $p \nmid a$, and if $y^2 \equiv a$ (mod p) is solvable, then there are exactly two elements in a complete solution, the other being congruent to $-y$. Thus if we compute

$$1^2, 2^2, \dots, (p-1)^2$$

we hit each congruence class (mod p) containing quadratic residues exactly twice, first with some k, $1 \le k \le (p-1)/2$, and the second time with $p - k$, since $(p - k)^2 \equiv k^2$ (mod p). Thus we have the following theorem, which at least cuts down by half the task of looking for solutions to $y^2 \equiv a$ (mod p).

THEOREM   Let p be any odd prime. Then any reduced residue system (mod p) contains $(p-1)/2$ quadratic residues and $(p-1)/2$ quadratic nonresidues (mod p). One set of $(p-1)/2$ incongruent quadratic residues is

$$1^2, 2^2, \dots, \left(\frac{p-1}{2}\right)^2.$$

Example   We compute the least residues (mod 11) of k^2, $k = 1, 2, \dots, 10$, in the following table.

k	1	2	3	4	5	6	7	8	9	10
$k^2 \equiv$	1	4	9	5	3	3	5	9	4	1

Note that each quadratic residue appears once between $k = 1$ and $5 = (11 - 1)/2$.

Example   Solve $2x^2 + 3x + 5 \equiv 0$ (mod 23).
   Here $b^2 - 4ac = 9 - 4(2)(5) = -31 \equiv 15$ (mod 23). We compute the least residues (mod 23) of the squares of the integers from 1 to $(23 - 1)/2 = 11$ to get the following table.

k	1	2	3	4	5	6	7	8	9	10	11
$k^2 \equiv$	1	4	9	16	2	13	3	18	12	8	6

Letting k run from 12 to 22 would produce the same squares in reverse order. (Try it!) Thus 15 is a quadratic nonresidue (mod 23) and the original congruence has no solution.

A Variation on Wilson's Theorem

The reader may want to review the proof of Wilson's Theorem, in which it is shown that $(p - 1)! \equiv -1 \pmod{p}$ by matching each of the integers k from $1, 2, \ldots, p - 1$ with the integer k' from that set such that $kk' \equiv 1 \pmod{p}$. We vary this proof as follows. Suppose p is an odd prime not dividing a, and match each k between 1 and $p - 1$ with a k' in that set such that $kk' \equiv a \pmod{p}$. Exactly one such k exists by Theorem 4.3.

There is a problem with the count if we could have $k = k'$, which says $k^2 \equiv a \pmod{p}$. This problem cannot arise if a is a quadratic nonresidue $\pmod{p}$, so let us assume that such is the case for the time being. Then we have $(p - 1)/2$ pairs kk', each with $kk' \equiv a \pmod{p}$. Thus

$$a^{(p-1)/2} \equiv 1 \cdot 2 \cdots (p - 1) \equiv -1 \pmod{p},$$

where of course the last congruence is Wilson's Theorem. We see that if a is a quadratic nonresidue $\pmod{p}$, then $a^{(p-1)/2} \equiv -1 \pmod{p}$.

If a is a quadratic residue $\pmod{p}$ we take a more direct approach. Let $y^2 \equiv a \pmod{p}$. Since $p \nmid a$, also $p \nmid y$. Then

$$a^{(p-1)/2} \equiv y^{p-1} \equiv 1 \pmod{p}$$

by Fermat's Theorem.

EULER'S CRITERION   Let p be an odd prime not dividing a. Then a is a quadratic residue or quadratic nonresidue $\pmod{p}$ according as

$$a^{(p-1)/2} \equiv 1 \text{ or } -1 \pmod{p}.$$

Unfortunately this theorem is not particularly practical for telling whether a number is a quadratic residue or not. In the last example the question was whether 15 was a quadratic residue $\pmod{23}$. Here $(p - 1)/2 = 11$, and so applying the theorem would entail computing the least residue of $15^{11} \pmod{23}$. This is not really easier than computing the residues of the squares from 1 to 11 as we did, and has the disadvantage of not giving a solution to $y^2 \equiv a \pmod{p}$ when there is one. Of course if powers of a are easily computed, as for $a = -1$, then the story changes.

COROLLARY   The number -1 is a quadratic residue or quadratic nonresidue modulo the odd prime p according as p is congruent to 1 or 3 $\pmod{4}$.

Proof   Of course any odd number p must be congruent to 1 or 3 $\pmod{4}$, and it is easy to check that then $(p - 1)/2$ is even or odd, respectively.

DEFINITION   the Legendre symbol

Suppose p is an odd prime not dividing a. We define the symbol (a/p) to be 1 or -1 according as a is a quadratic residue or quadratic nonresidue (mod p). This is called the **Legendre symbol**, after the French mathematician Adrien Marie Legendre (1752–1833).

Examples   From our previous computations we see that $(1/13) = (4/13) = (12/13) = 1$, while $(2/5) = (15/23) = -1$.

THEOREM 5.3   Let p be an odd prime not dividing a or b.

a. $\left(\dfrac{a^2}{p}\right) = 1.$

b. $\left(\dfrac{ab}{p}\right) = \left(\dfrac{a}{p}\right)\left(\dfrac{b}{p}\right).$

c. $\left(\dfrac{a}{p}\right) \equiv a^{(p-1)/2} \pmod{p}.$

d. If $a \equiv b \pmod{p}$,   then $\left(\dfrac{a}{p}\right) = \left(\dfrac{b}{p}\right).$

e. $\left(\dfrac{-1}{p}\right) = 1$ or -1 according as $p \equiv 1$ or $3 \pmod{4}.$

Proof   Parts (a) and (d) follow directly from the definition of quadratic residue. Part (c) is Euler's Criterion, and (e) is its corollary. Finally, part (b) follows from (c) since

$$\left(\frac{a}{p}\right)\left(\frac{b}{p}\right) \equiv a^{(p-1)/2}b^{(p-1)/2} = (ab)^{(p-1)/2} \equiv \left(\frac{ab}{p}\right) \pmod{p}.$$

PROBLEMS FOR SECTION 5.3

 In the first four problems list all the quadratic residues m (mod p), $0 < m < p$.

1. $p = 7$ 2. $p = 13$ 3. $p = 17$ 4. $p = 19$

In the next fourteen problems evaluate the given Legendre symbol.

5. $(3/5)$ 6. $(3/13)$ 7. $(101/3)$ 8. $(99/5)$
9. $(-10/13)$ 10. $(-2/13)$ 11. $(64/43)$ 12. $(121/47)$
13. $(-1/53)$ 14. $(-1/71)$ 15. $(72/73)$ 16. $(96/97)$
17. $(80/41)$ 18. $(234/47)$

In the next six problems find the least complete solution to the congruence. Start by making a table of the least residues of $1^2, 2^2, \ldots, 8^2$ (mod 17).

19. $2x^2 + 5x - 7 \equiv 0 \pmod{17}$ 20. $4x^2 - 7x + 2 \equiv 0 \pmod{17}$
21. $3x^2 + 4x + 5 \equiv 0 \pmod{17}$ 22. $3x^2 + 10x - 7 \equiv 0 \pmod{17}$
23. $x^2 + 9x + 3 \equiv 0 \pmod{17}$ 24. $5x^2 + 11x - 2 \equiv 0 \pmod{17}$

Tell how many elements a complete solution to the congruence has in each of the next six problems.

25. $x^2 + 5x + 7 \equiv 0 \pmod{19}$ 26. $x^2 + 7x + 2 \equiv 0 \pmod{41}$
27. $x^2 + 8x - 2 \equiv 0 \pmod{73}$ 28. $x^2 + 3x - 5 \equiv 0 \pmod{29}$
29. $x^2 + 9x - 1 \equiv 0 \pmod{43}$ 30. $x^2 + 9x + 1 \equiv 0 \pmod{41}$

In the next six problems find the least complete solution.

31. $4x^2 + 6x + 1 \equiv 0 \pmod{19}$ 32. $4x^2 + 6x + 1 \equiv 0 \pmod{11}$
33. $4x^2 + 6x + 1 \equiv 0 \pmod{13}$ 34. $x^2 + 16x + 32 \equiv 0 \pmod{127}$
35. $5x^2 + 8x + 1 \equiv 0 \pmod{129}$ 36. $3x^2 + 3x - 5 \equiv 0 \pmod{53}$

B

37. Finish the proof of the first theorem of this section by showing that if p is an odd prime not dividing a, if $y^2 \equiv b^2 - 4ac \pmod{p}$, and if $2ax \equiv -b + y \pmod{p}$, then $ax^2 + bx + c \equiv 0 \pmod{p}$. (**Hint:** Find u such that $2au \equiv 1 \pmod{p}$. Then $x \equiv u(-b + y) \pmod{p}$. Substitute this into the desired congruence.)

38. Show that if $m > 0$ and if a and b are integers such that $x^2 \equiv a \pmod{m}$ and $y^2 \equiv b \pmod{m}$ are both solvable, then $z^2 \equiv ab \pmod{m}$ is solvable.

39. Show that if p is an odd prime and $p \nmid ac$, then complete solutions to $ax^2 + bx + c \equiv 0 \pmod{p}$ and $cx^2 + bx + a \equiv 0 \pmod{p}$ have the same number of solutions.

40. Suppose p is an odd prime not dividing ab. Show that $z^2 \equiv ab \pmod{p}$ is solvable exactly when both or neither of $x^2 \equiv a \pmod{p}$ and $y^2 \equiv b \pmod{p}$ are solvable.

In the next four problems tell whether the statement is true and give a counter-example if it is false.

41. Let $m > 0$ and $(m, ab) = 1$. If neither $x^2 \equiv a \pmod{m}$ nor $y^2 \equiv b \pmod{m}$ is solvable, then $z^2 \equiv ab \pmod{m}$ is solvable.

42. If p is an odd prime then complete solutions to $ax^2 + bx + c \equiv 0 \pmod{p}$ and $cx^2 + bx + a \equiv 0 \pmod{p}$ have the same number of elements.

43. If p is a prime congruent to 3 (mod 4) and $p \nmid a$, then exactly one of a and $-a$ is a quadratic residue (mod p).

44. If p is prime and $p \nmid a$, then the number of elements in a complete solution to $x^2 \equiv a \pmod{p^n}$ is 0 or 2.

C

45. Show that if $m > 0$ is the product of distinct odd primes and $(m, a) = 1$, then the number of elements in a complete solution to $x^2 \equiv a \pmod{m}$ is 0 or a power of 2.

46. Show that if p is an odd prime not dividing a, then a complete solution to $x^2 \equiv a \pmod{p^n}$ has the same number of elements for all positive integers n.
47. Show that if m is odd and $(m, a) = 1$, then the number of elements in a complete solution to $x^2 \equiv a \pmod{m}$ is 0 or a power of 2.

5.4   Quadratic Reciprocity

Identifying Quadratic Residues

The problem of telling whether an integer a is a quadratic residue (mod p) or not occupied some of the greatest number theorists of the eighteenth century, including Euler, Lagrange, Legendre, and Gauss, who made the greatest contribution to the subject. To save words we will establish the following convention:

CONVENTION   In this section p will represent an odd prime not dividing the positive integer a. All congruences will be modulo p unless some other modulus is specified.

Note that we can assume a to be positive without loss of generality, since by part (b) of Theorem 5.3 $(-a/p) = (-1/p)(a/p)$ and $(-1/p)$ can be evaluated by part (e) of the same theorem.

About all we have to go on is the identity

$$\left(\frac{a}{p}\right) \equiv a^{(p-1)/2},$$

and it would be nice to have some independent way to evaluate the expression on the right. In Fermat's Theorem we evaluated a similar expression, showing that

$$a^{p-1} \equiv 1.$$

Recall the proof. As x runs through a reduced residue system (mod p) so does ax, so

$$a^{p-1}\Pi x = \Pi(ax) \equiv \Pi x,$$

where the products run over $x = 1, 2, \ldots, p - 1$. Dividing through by Πx yields Fermat's Theorem.

In order to evaluate $a^{(p-1)/2}$ we need to employ a product with $(p - 1)/2$ factors, say over $x = 1, 2, \ldots, (p-1)/2$. Let us define the integer h (think of "half") to be $(p - 1)/2$. The integers from 1 to $p - 1$ then look like

$$1, 2, \ldots, \frac{p-1}{2} = h, h + 1 = \frac{p+1}{2}, \ldots, p - 1.$$

A proof like that for Fermat's Theorem would have us look at the product

$$a^h \Pi x = \Pi(ax),$$

where x runs from 1 to h in the products.

The problem with this product is that for x between 1 and h, ax need not be congruent to some number in the same range. Certainly ax is congruent to one of the numbers

$$-h, \; -(h-1), \ldots, \; -1, 0, 1, 2, \ldots, h,$$

since these $2h + 1 = p$ integers comprise a complete residue system (mod p). In fact we could leave out 0 as a possibility, since by our assumptions $p \nmid ax$.

Let us define x^* to be that unique number between $-h$ and h such that $ax \equiv x^*$, where x runs from 1 to h. Taking $p = 7$ and $a = 5$, for example, we compute the following table. Note that $h = (7-1)/2 = 3$.

x	1	2	3
ax	5	10	15
x^*	-2	3	1

Here is another example, with $p = 13$ (so $h = 6$) and $a = 2$.

x	1	2	3	4	5	6
ax	2	4	6	8	10	12
x^*	2	4	6	-5	-3	-1

Notice that the values of x^* repeat those of x, except for some minus signs. This turns out always to be the case, which is the basis of the following theorem.

GAUSS'S LEMMA   Let p be an odd prime not dividing a. For $x = 1, 2, \ldots, h = (p-1)/2$, let x^* be that unique integer congruent to ax (mod p) such that $-h \le x^* \le h$. Suppose exactly n values of x^* are negative. Then $(a/p) = (-1)^n$.

Proof   First we show that the absolute values of the numbers x^* comprise the set $\{1, 2, \ldots, h\}$. Since these absolute values all fall into this set, and since there are exactly h of them, it suffices to show they are distinct, that is, if $x \ne y$, then $|x^*| \ne |y^*|$.

Suppose $|x^*| = |y^*|$, with x and y between 1 and h. If x^* and y^* have the same sign, then $x^* = y^*$, and so $ax \equiv ay$ and $x = y$ by the Cancellation Theorem. Otherwise $x^* = -y^*$, which means $ax \equiv -ay$. Thus $p \mid a(x+y)$. This is impossible because x and y are between 1 and h, so $2 \le x + y \le 2h = p - 1$. Thus p cannot divide $x + y$, and by assumption p doesn't divide a.

The rest of the proof parallels that of Fermat's Theorem. If we let x run from 1 to h in the products below, then by the definition of n

$$a^h \Pi x = \Pi(ax) \equiv \Pi x^* \equiv (-1)^n \Pi x.$$

Canceling Πx gives $(-1)^n \equiv a^h$. But the latter is congruent to (a/p) by Theorem 5.3(c). Since $p > 2$ we must have $(-1)^n = (a/p)$.

The Case $a = 2$

We will use Gauss's Lemma to compute (a/p) for some specific values of a. It is important to remember that we don't really need to know exactly what n is, only whether it is odd or even. (This is called the **parity** of n.)

Let us see what happens when $a = 2$. We are to find numbers x^* congruent to the numbers

$$2, 4, 6, \ldots, 2h = p - 1$$

in the set $\{-h, -h + 1, \ldots, h\}$, and count how many of them are negative. (One example of how this works is given for $p = 13$ earlier in this section.) In general n is the number of integers x such that $p/2 < 2x < p$, as the following picture shows.

Thus we must count the number of integers x such $p/4 < x < p/2$.

Of course $p/4$ is not even an integer (nor is $p/2$), and so the smallest such x depends on what p is congruent to (mod 4). Let us suppose $p = 4q + r, 0 \leq r < 4$. Since p is odd we must have $r = 1$ or 3. Then we must count the integers x such that

$$\frac{p}{4} = q + \frac{r}{4} < x < 2q + \frac{r}{2} = \frac{p}{2}.$$

The first x counted is clearly $q + 1$, and the last one counted is $2q$ or $2q + 1$, depending on whether $r = 1$ or 3. Thus $n = q$ or $q + 1$, according as $r = 1$ or 3.

The question is whether n is even or odd, which we see depends on both q and r. Let us check cases.

case	$p = 4q + r$	n	
$r = 1, q = 2s$	$8s + 1$	$q,$	even
$r = 1, q = 2s + 1$	$8s + 5$	$q,$	odd
$r = 3, q = 2s$	$8s + 3$	$q + 1,$	odd
$r = 3, q = 2s + 1$	$8s + 7$	$q + 1,$	even

THEOREM 5.4 Let p be an odd prime. Then $(2/p)$ is 1 if $p \equiv 1$ or 7 (mod 8), and $(2/p)$ is -1 if $p \equiv 3$ or 5 (mod 8).

Example Compute $\left(\dfrac{65}{47}\right)$.

Using various parts of Theorem 5.3 and the result just proved we have

$$\left(\frac{65}{47}\right) = \left(\frac{18}{47}\right) = \left(\frac{2}{47}\right)\left(\frac{9}{47}\right) = \left(\frac{2}{47}\right) = 1,$$

where the last equality holds since $47 \equiv 7$ (mod 8).

The Case $a = 3$

A similar argument to the one above can be made for $a = 3$. Since we are assuming $p \nmid a$, then $p > 3$. The multiples of $1, \dots, h$ are

$$3, 6, 9, \dots, 3h = \frac{3(p-1)}{2},$$

all of which lie between 0 and $3p/2$. The following table shows what happens for $p = 17$ (so $h = 8$).

x	1	2	3	4	5	6	7	8
$3x$	3	6	9	12	15	18	21	24
x^*	3	6	-8	-5	-2	1	4	7

The negative values of x^* correspond to $p/2 < 3x < p$, which is equivalent to $p/6 < x < p/3$.

The first integer after $p/6$ depends on the least residue of p (mod 6), but in light of our experience with $a = 2$ we will use the modulus 12 instead. Let $p = 12s + r$, $0 \le r < 12$. Since neither 2 nor 3 divides p, we must have $r = 1, 5, 7,$ or 11.

Substituting $p = 12s + r$ into our inequality for x gives

$$2s + \frac{r}{6} < x < 4s + \frac{r}{3}.$$

Since shifting either endpoint of an interval by an even integer does not change whether the number of integers in the interval is even or odd, it suffices to count the number of integers y satisfying

$$\frac{r}{6} < y < \frac{r}{3}.$$

The number of such y will be even or odd the same as n. We resort to cases:

case	interval	possible y	parity of n
$r = 1$	$\frac{1}{6} < y < \frac{1}{3}$	none	even
$r = 5$	$\frac{5}{6} < y < 1\frac{1}{3}$	1	odd
$r = 7$	$1\frac{1}{6} < y < 2\frac{1}{3}$	2	odd
$r = 11$	$1\frac{5}{6} < y < 3\frac{2}{3}$	2, 3	even

THEOREM Let $p > 3$ be prime. Then $(3/p) = 1$ if $p \equiv 1$ or 11 (mod 12) and $(3/p) = -1$ if $p \equiv 5$ or 7 (mod 12).

Quadratic Reciprocity

The value of $(2/p)$ depends on what p is (mod 8). The value of $(3/p)$ depends on what p is (mod 12). Furthermore, some symmetry seems present. For example $(2/p)$ has the same value whether $p \equiv 1$ (mod 8) or $p \equiv -1 \equiv 7$ (mod 8), and the same value whether $p \equiv 3$ (mod 8) or $p \equiv -3 \equiv 5$ (mod 8). The situation is similar for $a = 3$. Thus we might conjecture the following theorem.

THEOREM Q Let $a > 0$, and let p and q be odd primes not dividing a. Then $(a/p) = (a/q)$ if $p \equiv q$ (mod $4a$) or if $p \equiv -q$ (mod $4a$).

The proof of this theorem, which proceeds along lines similar to the cases $a = 2$ and $a = 3$ above, will be presented in Section 5.5. For the time being we will assume it is true, and investigate its consequences. Suppose p and q are distinct odd primes, so that each is congruent to either 1 or 3 (mod 4). If $p \equiv q$ (mod 4), then, assuming $p < q$, we have $q - p = 4a$ for some positive integer a. Then $p \equiv q$ (mod $4a$), and so $(a/p) = (a/q)$ by Theorem Q. Since $q - p = 4a$, the integer $4a$ has the interesting property that

$$q \equiv 4a \quad (\text{mod } p) \qquad \text{and} \qquad p \equiv -4a \quad (\text{mod } q).$$

Thus

$$\left(\frac{p}{q}\right) = \left(\frac{-4a}{q}\right) = \left(\frac{-1}{q}\right)\left(\frac{4}{q}\right)\left(\frac{a}{q}\right) = \left(\frac{-1}{q}\right)\left(\frac{4}{p}\right)\left(\frac{a}{p}\right)$$

$$= \left(\frac{-1}{q}\right)\left(\frac{4a}{p}\right) = \left(\frac{-1}{q}\right)\left(\frac{q}{p}\right),$$

where this computation is justified by various parts of Theorem 5.3 (including $(4/q) = (4/p) = 1$ by part (a)) along with the equation $(a/p) = (a/q)$.

Recall that we know that $(-1/q)$ is 1 or -1 according as q (and so p also) is congruent to 1 or 3 (mod 4). We see that (p/q) and (q/p) are equal if both p

and q are congruent to 1 (mod 4), but $(p/q) = -(q/p)$ if both are congruent to 3 (mod 4).

If $p \not\equiv q$ (mod 4) the situation is even simpler. Then $p + q = 4a$ for some positive integer a. Since $p \equiv -q$ (mod $4a$), Theorem Q says that $(a/p) = (a/q)$. Now we have

$$p \equiv 4a \quad (\text{mod } q) \qquad \text{and} \qquad q \equiv 4a \quad (\text{mod } p),$$

and so

$$\left(\frac{p}{q}\right) = \left(\frac{4a}{q}\right) = \left(\frac{4}{q}\right)\left(\frac{a}{q}\right) = \left(\frac{4}{p}\right)\left(\frac{a}{p}\right) = \left(\frac{4a}{p}\right) = \left(\frac{q}{p}\right).$$

We have proved the following celebrated theorem.

THE LAW OF QUADRATIC RECIPROCITY (Gauss)   Let p and q be distinct odd primes. Then $(p/q) = (q/p)$ unless p and q are both congruent to 3 (mod 4), in which case $(p/q) = -(q/p)$.

Examples   Evaluate $(13/43)$, $(19/59)$, and $(37/67)$.
Since $13 \equiv 1$ (mod 4) we have

$$\left(\frac{13}{43}\right) = \left(\frac{43}{13}\right) = \left(\frac{3 \cdot 13 + 4}{13}\right) = \left(\frac{4}{13}\right) = 1,$$

where we used, besides reciprocity, parts (d) and (a) of Theorem 5.3.
Likewise since both 19 and 59 are congruent to 3 (mod 4) we have

$$\left(\frac{19}{59}\right) = -\left(\frac{59}{19}\right) = -\left(\frac{2}{19}\right) = 1,$$

since $(2/19) = -1$ by Theorem 5.4.
Finally since $37 \equiv 1$ (mod 4) we have

$$\left(\frac{37}{67}\right) = \left(\frac{67}{37}\right) = \left(\frac{30}{37}\right) = \left(\frac{2}{37}\right)\left(\frac{3}{37}\right)\left(\frac{5}{37}\right) = (-1)\left(\frac{37}{3}\right)\left(\frac{37}{5}\right)$$

$$= -\left(\frac{1}{3}\right)\left(\frac{2}{5}\right) = -(1)(-1) = 1,$$

where Theorem 5.4 was used twice. Alternatively, we could note that $67 \equiv -7$ (mod 37), so

$$\left(\frac{67}{37}\right) = \left(\frac{-7}{37}\right) = \left(\frac{-1}{37}\right)\left(\frac{37}{7}\right) = (1)\left(\frac{2}{7}\right) = 1.$$

PROBLEMS FOR SECTION 5.4

 In the first six problems use Theorems 5.3 and 5.4 to evaluate the given Legendre symbol.

1. (32/101)   2. (50/71)   3. (228/113)
4. (115/97)   5. (66/67)   6. (65/67)

In the next six problems use Theorems 5.3 and 5.4 and the theorem in this section concerning (3/p) to evaluate the given Legendre symbol, if possible.

7. (3/83)   8. (−3/89)   9. (77/71)
10. (600/131)   11. (27/91)   12. (162/53)

In the next ten problems evaluate the given Legendre symbol, using the Law of Quadratic Reciprocity if necessary.

13. (17/67)   14. (19/67)   15. (53/73)   16. (86/103)
17. (1000/107)   18. (−1066/83)   19. (17/37)   20. (23/43)
21. (44/71)   22. (71/101)

In the next eight problems determine whether the congruence has a solution.

23. $x^2 \equiv 15 \pmod{41}$ 24. $x^2 \equiv 1000 \pmod{157}$
25. $x^2 \equiv 234 \pmod{157}$ 26. $x^2 \equiv 158 \pmod{159}$
27. $x^2 - 3x + 17 \equiv 0 \pmod{61}$ 28. $x^2 + 5x + 7 \equiv 0 \pmod{97}$
29. $3x^2 + 4x + 5 \equiv 0 \pmod{51}$ 30. $4x^2 + 5x + 6 \equiv 0 \pmod{87}$

 In the next four problems tell the number of elements in a complete solution to the given congruence.

31. $x^2 \equiv 17 \pmod{143}$ 32. $x^2 \equiv 23 \pmod{119}$
33. $3x^2 + 7x - 2 \equiv 0 \pmod{97}$ 34. $2x^2 + 7x - 5 \equiv 0 \pmod{1099}$

35. Show that $(2/p) = (-1)^{(p^2-1)/8}$ for any odd prime p.
36. Show that the Law of Quadratic Reciprocity is equivalent to the statement: If p and q are distinct odd primes, then $(p/q)(q/p) = (-1)^{(p-1)(q-1)/4}$.
37. Use the Law of Quadratic Reciprocity to show that if p is an odd prime other than 5, then $(5/p) = 1$ if and only if the last digit of p is 1 or 9.
38. Make a table of $x = 1, 2, \ldots, h$ and x^* for $p = 17$ and $a = 5$.
39. Repeat the last problem for $p = 19$ and $a = 5$.
40. Show that if p is an odd prime other than 5 and r is the least residue of p (mod 20), then r is 1, 3, 7, 9, 11, 13, 17, or 19.
41. Find the number of integers y satisfying $r/10 < y < r/5$ for each r in the previous problem.
42. Find the number of integers y satisfying $3r/10 < y < 2r/5$ for each r in the second previous problem.

C

In the next three problems evaluate (5/p), where p is an odd prime other than 5, in a manner like that used for (2/p) and (3/p) in this section. Do not use the Law of Quadratic Reciprocity in these problems. Some of the work is already done in the previous three problems.

43. Show that if h and x^* are defined as in this section, then the number of x^* that are negative for $x = 1, 2, \ldots, h$ is the number n of integers x satisfying $p/10 < x < p/5$ or $3p/10 < x < 2p/5$.
44. Show that if r is the least residue of p (mod 20), then the number n of the previous problem is even or odd according as the number of integers y satisfying either $r/10 < y < r/5$ or $3r/10 < y < 2r/5$ is even or odd.
45. Show that $(5/p)$ is 1 if $p \equiv 1, 9, 11,$ or 19 (mod 20), and is -1 if $p \equiv 3, 7, 13,$ or 17 (mod 20).

46. Show that the last problem agrees with the Law of Quadratic Reciprocity.
47. Show that if $4a \mid p - q$, where p and q are odd primes, then $(a/p) = (a/q)$ even if a is negative.
48. Show that if $ab \equiv 1$ (mod p), p an odd prime, then $(a/p) = (b/p)$.

5.5   The Jacobi Symbol

The Proof of Theorem Q

Our first order of business is to prove the theorem of the last section from which we derived the Law of Quadratic Reciprocity.

THEOREM Q   Let p and q be odd primes not dividing $a > 0$. Then $(a/p) = (a/q)$ if $p \equiv q$ (mod $4a$) or if $p \equiv -q$ (mod $4a$).

Proof   As with our evaluations of $(2/p)$ and $(3/p)$, we will employ Gauss's Lemma. Let $h = (p - 1)/2$, and consider the integers

$$a, 2a, 3a, \ldots, ha.$$

These fall into the open intervals

$$\left(0, \frac{p}{2}\right), \left(\frac{p}{2}, \frac{2p}{2}\right), \left(\frac{2p}{2}, \frac{3p}{2}\right), \left(\frac{3p}{2}, \frac{4p}{2}\right), \ldots, \tag{1}$$

where, as is customary, we are denoting the set of real numbers X such that $A < X < B$ by the notation (A, B). Since

$$ha = \frac{(p-1)a}{2} < \frac{ap}{2} < \frac{(p+1)a}{2} = (h+1)a,$$

the last interval we need consider is $((a-1)p/2, ap/2)$. A total of a intervals are involved, so that the number of intervals does not depend on p.

Notice that the endpoints of the intervals listed in (1) are either nonintegers or else multiples of p. Thus none of the integers $a, 2a, \ldots, ha$ falls on one of these endpoints, since $p \nmid a$ and $h < p$.

As in Section 5.4 we define x^* by $x^* \equiv ax \pmod{p}$, $-h \le x^* \le h$. By Gauss's Lemma the value of (a/p) depends on whether the number of negative x^* is even or odd. The integer x^* will be negative when ax falls in half the intervals listed in (1), namely the intervals

$$\left(\frac{p}{2}, \frac{2p}{2}\right), \left(\frac{3p}{2}, \frac{4p}{2}\right), \left(\frac{5p}{2}, \frac{6p}{2}\right), \ldots.$$

Thus in a typical interval we want to count the number of integers x such that

$$\frac{(2k-1)p}{2} < ax < \frac{2kp}{2},$$

or

$$\frac{(2k-1)p}{2a} < x < \frac{2kp}{2a}. \tag{2}$$

Now assume q is an odd prime such that $q \equiv p \pmod{4a}$. Then $q = p + 4at$ for some integer t. If we try to evaluate (a/q) by Gauss's Lemma in the same way, a typical interval in which we would be counting integers would be defined by the inequalities

$$\frac{(2k-1)q}{2a} < y < \frac{2kq}{2a}. \tag{3}$$

Plugging $q = p + 4at$ into this leads to

$$\frac{(2k-1)p}{2a} + (2k-1)2t < y < \frac{2kp}{2a} + 4kt. \tag{4}$$

(We leave it to the reader to check the algebra.)

If we compare the endpoints of the intervals defined by the inequalities (2) and (4) we see that the left endpoints differ by even integers, as do the right endpoints. Thus the number of integers in the corresponding intervals differs by a multiple of 2; it is even in both cases or odd in both cases. By using the same argument for each value of k and applying Gauss's Lemma we conclude that $(a/p) = (a/q)$.

Now we consider the case when $q \equiv -p \pmod{4a}$. Then $q = -p + 4at$ for some integer t. Plugging this into (3) produces

$$\frac{-(2k-1)p}{2a} + (2k-1)2t < y < \frac{-2kp}{2a} + 4kt.$$

Multiplying through by -1 produces a symmetric interval on the other side of 0 that contains the same number of integers:

$$\frac{(2k-1)p}{2a} - (2k-1)2t > y' > \frac{2kp}{2a} - 4kt.$$

In fact the same number of integers are in the interval shifted $4kt$ units to the right:

$$\frac{(2k-1)p}{2a} + 2t > y'' > \frac{2kp}{2a},$$

which can be written

$$\frac{2kp}{2a} < y'' < \frac{(2k-1)p}{2a} + 2t. \tag{5}$$

We would like to show that the number of integers y'' satisfying these inequalities is even or odd the same as the number of x satisfying (2). But (5) and (2) define adjacent intervals, and the number of integers in their union is the number of z satisfying

$$\frac{(2k-1)p}{2a} < z < \frac{(2k-1)p}{2a} + 2t.$$

(Recall that the endpoints of our intervals are never hit so we need not worry about x equaling the common endpoint of the two intervals.)

The last inequalities define an interval of length $2t$ with nonintegral endpoints. It must contain an even number of integers. Thus the number of integers satisfying (2) and (5) must be even in both cases or odd in both cases. Again, using this argument for all values of k and applying Gauss's Lemma we see that $(a/q) = (a/p)$.

A Generalization of the Legendre Symbol

Calculation with the Legendre symbol (a/p) is hampered by the fact that a must be prime to use the reciprocity law. To determine $(105/211) = (3 \cdot 5 \cdot 7/211)$, for example, we must invert (3/211), (5/211), and (7/211) separately. We know that

$$\left(\frac{105}{211}\right) = \pm\left(\frac{211}{3}\right)\left(\frac{211}{5}\right)\left(\frac{211}{7}\right);$$

let us see if we can discover some short way to tell whether the plus or minus sign is correct.

Suppose q is an odd prime not dividing the odd number m, and that $m = p_1 p_2 \cdots p_t$. We are looking at

$$\left(\frac{m}{q}\right) = \left(\frac{p_1 p_2 \cdots p_t}{q}\right) = \left(\frac{p_1}{q}\right)\left(\frac{p_2}{q}\right)\cdots\left(\frac{p_t}{q}\right).$$

If $q \equiv 1 \pmod 4$ everything is easy, for then $(p_i/q) = (q/p_i)$ for all i. Let us suppose $q \equiv 3 \pmod 4$. Then

$$\left(\frac{p_1 p_2 \cdots p_t}{q}\right) = (-1)^n \left(\frac{q}{p_1}\right)\left(\frac{q}{p_2}\right)\cdots\left(\frac{q}{p_t}\right),$$

where n is just the number of p's congruent to $-1 \pmod 4$. But since each p is congruent to $\pm 1 \pmod 4$, we have $m = p_1 \cdots p_t \equiv (-1)^n \pmod 4$, where n is exactly the number mentioned in the last sentence.

We see that $(m/q) = (q/p_1)(q/p_2)\cdots(q/p_t)$ unless m and q are both congruent to 3 (mod 4), in which case $(m/q) = -(q/p_1)(q/p_2)\cdots(q/p_t)$. This sounds exactly like the Quadratic Reciprocity Law, except that we must write $(q/p_1)(q/p_2)\cdots(q/p_t)$ instead of (q/m), since the latter is undefined. An enlargement of our definition is indicated.

DEFINITION   the Jacobi symbol

Suppose $m = p_1 p_2 \cdots p_t$, where the p's are odd primes, and that $(a, m) = 1$. We define (a/m), called the **Jacobi symbol**, by

$$\left(\frac{a}{m}\right) = \left(\frac{a}{p_1}\right)\left(\frac{a}{p_2}\right)\cdots\left(\frac{a}{p_t}\right).$$

Notice that if m is itself prime then (a/m) has its previous meaning. Thus using the same notation as for the Legendre symbol is justified. Of course we cannot just assume all the results we proved for the Legendre symbol also hold for the Jacobi symbol. They must be checked out again.

THEOREM 5.5   Suppose m and n are relatively prime positive odd numbers. Then $(m/n) = (n/m)$ unless $m \equiv n \equiv 3 \pmod 4$, in which case $(m/n) = -(n/m)$.

Proof Let $m = p_1 \cdots p_t$ and $n = q_1 \cdots q_u$, the p's and q's prime. From the definition of the Jacobi symbol and part (b) of Theorem 5.3 we have

$$\left(\frac{m}{n}\right) = \prod_{j=1}^{u} \left(\frac{m}{q_j}\right) = \prod_{j=1}^{u} \prod_{i=1}^{t} \left(\frac{p_i}{q_j}\right).$$

Let r be the number of i, $1 \le i \le t$, such that $p_i \equiv 3 \pmod 4$, and let s be the number of j, $1 \le j \le u$, such that $q_j \equiv 3 \pmod 4$. Then the number of pairs i, j such that $p_i \equiv q_j \equiv 3 \pmod 4$ is rs. Thus from the Law of Quadratic Reciprocity we see

$$\prod_{j=1}^{u} \prod_{i=1}^{t} \left(\frac{p_i}{q_j}\right) = (-1)^{rs} \prod_{j=1}^{u} \prod_{i=1}^{t} \left(\frac{q_j}{p_i}\right) = (-1)^{rs} \left(\frac{n}{m}\right).$$

The argument before the definition of the Jacobi symbol shows $m \equiv 3 \pmod 4$ if and only if r is odd. In the same way $n \equiv 3 \pmod 4$ if and only if s is odd. Since $(-1)^{rs} = 1$ unless r and s are both odd, the theorem follows.

THEOREM 5.6 Suppose m and n are odd positive integers, and $(a, m) = (b, m) = (a, n) = 1$.

a. $\left(\dfrac{a}{mn}\right) = \left(\dfrac{a}{m}\right)\left(\dfrac{a}{n}\right).$

b. $\left(\dfrac{ab}{m}\right) = \left(\dfrac{a}{m}\right)\left(\dfrac{b}{m}\right).$

c. If $a \equiv b \pmod m$, then $\left(\dfrac{a}{m}\right) = \left(\dfrac{b}{m}\right).$

d. $\left(\dfrac{-1}{m}\right) = 1$ if and only if $m \equiv 1 \pmod 4$.

e. $\left(\dfrac{2}{m}\right) = 1$ if and only if $m \equiv \pm 1 \pmod 8$.

We will leave the proof of this theorem, which depends on the definition of the Jacobi symbol and the corresponding theorems concerning the Legendre symbol, for the exercises.

Example Evaluate $(105/211)$.
Since $105 \equiv 1 \pmod 4$ we have

$$\left(\frac{105}{211}\right) = \left(\frac{211}{105}\right) = \left(\frac{1}{105}\right) = 1.$$

Note that it is not even necessary to check out whether 105 and 211 are prime.

In order to draw the conclusion that $x^2 \equiv 105 \pmod{211}$ has a solution, however, it is necessary to know that 211 is prime. (It is.)

WARNING: If m is not prime, it may be that $(a/m) = 1$ even though $x^2 \equiv a \pmod{m}$ has no solutions.

Example We have $(2/15) = (2/3)(2/5) = (-1)(-1) = 1$, yet $x^2 \equiv 2 \pmod{15}$ has no solution. (How could it have, when $x^2 \equiv 2 \pmod 3$ has no solution?)

In spite of the above warning, the Jacobi symbol can be useful in the determination of quadratic residues. The symbol $(211/105)$, which appeared in the first example after Theorem 5.6, is not a Legendre symbol. Yet the conclusion that 105 is a quadratic residue $\pmod{211}$ is perfectly valid.

Example Evaluate $(10/77)$.
We have

$$\left(\frac{10}{77}\right) = \left(\frac{2}{77}\right)\left(\frac{5}{77}\right) \quad \text{[Theorem 5.3(b)]}$$

$$= (-1)\left(\frac{77}{5}\right) \quad \text{[Theorems 5.6(e), 5.5]}$$

$$= -\left(\frac{2}{5}\right) \quad \text{[Theorem 5.3(d)]}$$

$$= -(-1) = 1. \quad \text{[Theorem 5.4 or 5.6(e)]}$$

PROBLEMS FOR SECTION 5.5

 In the first ten problems name the earliest numbered theorem or part of a theorem that justifies the equation.

1. $(55/91) = (5/91)(11/91)$
2. $(51/59) = -(59/51)$
3. $(169/61) = 1$
4. $(2/97) = 1$
5. $(401/53) = (30/53)$
6. $(11/65) = (11/5)(11/13)$
7. $(2/111) = 1$
8. $(10/13) = (10/53)$
9. $(-1/47) = -1$
10. $(19/21) = (-2/21)$

In the next ten problems evaluate the given symbol, if possible.

11. $(77/97)$
12. $(210/451)$
13. $(128/315)$
14. $(361/612)$
15. $(-30/79)$
16. $(123/217)$
17. $(210/143)$
18. $(247/-433)$
19. $(-433/247)$
20. $(5041/53)$

In the next six problems tell whether the given congruence has solutions.

21. $x^2 \equiv 95 \pmod{101}$
22. $x^2 \equiv 101 \pmod{95}$
23. $x^2 \equiv 11 \pmod{1045}$
24. $x^2 \equiv 211 \pmod{1045}$
25. $x^2 \equiv 2 \pmod{1763}$
26. $x^2 \equiv -1 \pmod{1849}$

27. Prove part (a) of Theorem 5.6.   28. Prove part (b) of Theorem 5.6.
29. Prove part (c) of Theorem 5.6.   30. Prove part (d) of Theorem 5.6.
31. Show that in the proof of Theorem Q the maximum value of k in (2) is $a/2$ if a is even and $(a - 1)/2$ if a is odd.
32. Consider the proof of Theorem Q when $a = 7$, $p = 11$, and $q = 67$, so $p \equiv q$ (mod 4a). Count the number of integers satisfying (2) for $k = 1$, 2, and 3. Do the same for (3), and confirm that corresponding intervals both contain even or odd numbers of integers.
33. Repeat the previous problem, but change q to 73. Then $p \equiv -q$ (mod 4a), and in fact $p + q = 4at$ with $t = 3$. Count the integers satisfying (3) for $k = 1$, 2, and 3 and confirm that for each k the total number of integers satisfying (2) or (3) is $2t = 6$.
34. Suppose $p > 3$ is prime. Show that $(-3/p) = 1$ if and only if $p \equiv 1$ (mod 3).
35. Let p be an odd prime. When does $x^2 \equiv (2/p)$ (mod p) have no solutions?
36. Let m be odd and $(a, m) = 1$. Show that if $(a/m) = -1$, then $x^2 \equiv a$ (mod m) has no solutions.
37. Show that if m is an odd positive integer and $(a, m) = 1$, then $(a/m^2) = 1$.

38. Prove part (e) of Theorem 5.6. (**Hint:** Let $m = p_1 \cdots p_t$, the p's prime. Show that $m \equiv \pm 3$ (mod 8) if and only if $p_i \equiv \pm 3$ (mod 8) for an odd number of i.)
39. Let p be an odd prime other than 7. When is $x^2 \equiv (7/p)$ (mod p) without solutions?
40. Show that if F_r is the rth Fermat number, $r > 0$, then $(3/F_r) = -1$.

Joseph Louis Lagrange
1736–1813

Lagrange, born in Turin, Italy, was of French and Italian ancestry. He became interested in mathematics as a youth, studying it on his own with such success that he was appointed professor of geometry at the Royal Artillery School in Turin at the age of 16. Thus began a long career establishing him as

one of the two greatest mathematicians of the eighteenth century. (Euler was the other.)

Lagrange's modesty and nondogmatic manner allowed him to get along with diverse personalities in a contentious period of history. In 1766 he took the place of Euler, who was always generous in his praise of Lagrange's work, in the Berlin court of Frederick the Great of Prussia. When Frederick died 20 years later, Lagrange went to Paris at the invitation of Louis XVI as a member of the French Academy. Although he became a special favorite of Marie Antoinette, during the French Revolution he stayed in Paris and was given a pension and employment by the revolutionary regime, including the presidency of the commission that introduced the metric system. Later, Napoleon enjoyed conversations with him.

JOSEPH LOUIS LAGRANGE

Lagrange is mainly known as an analyst. He made an attempt to put calculus on a rigorous foundation, which, while unsuccessful, set goals that were achieved by later mathematicians. His is the notation $f', f'', \ldots$ for the derivatives of a function. In algebra he is responsible for the theorem that says the number of elements in a subgroup of a finite group always divides the number of elements in the group.

Although Lagrange thought of number theory mainly as a diversion, he proved many important arithmetic results. He showed that every positive integer can be written as the sum of four squares, a result Euler worked on but failed to prove. To Lagrange is due also the first proof of Wilson's Theorem, and the theorem that a congruence of degree n with prime modulus cannot have more than n solutions.

Chapter 6

The Number Theory of the Reals

Although there are many real numbers, rational and irrational, that are not integers, the integers may be important in studying these numbers. Connections between the integers and nonintegers will be investigated in this chapter.

6.1   Rational and Irrational Numbers

Countability

That not all real numbers are rational, that is, the quotient of two integers, was known to ancient Greek mathematicians. We looked at the usual proof that $\sqrt{2}$ is irrational back in Chapter 0. With what we know now about prime factorization, that proof can be generalized considerably. It turns out that $\sqrt[k]{n}$ is irrational whenever n is not a perfect kth power.

THEOREM   Let n and k be positive integers. Then if $\sqrt[k]{n}$ is not an integer, it is irrational.

Proof   Suppose that $\sqrt[k]{n}$ is rational, say a/b, where a and b are positive integers. Then

$$b^k n = a^k.$$

Let us consider the factorizations of a, b, and n into prime powers. Suppose the prime p occurs in the factorizations of a, b, and n to the powers r, s, and t, respectively. Since p appears the same number of times in the factorizations

185

of both sides of our equation we have

$$(p^s)^k p^t = (p^r)^k, \qquad sk + t = rk.$$

We see that t, the power of p occurring in n, must be a multiple of k. Since this holds for each prime dividing n, we can conclude that n is a perfect kth power, and so $\sqrt[k]{n}$ is an integer.

This theorem lets us exhibit as many irrational numbers as we wish, for example $\sqrt{3}, \sqrt{5}, \sqrt{6}, \sqrt[3]{3}$, and $\sqrt[13]{7}$. As a matter of fact there are more irrational numbers than rational, but some definitions need to be given to make this statement valid. There are infinitely many rational numbers and infinitely many irrational, so what is needed is a way of comparing infinite sets.

DEFINITIONS   countable, uncountable

We say a set S is **countable** if it can be put into a one-to-one correspondence with the set of positive integers. This simply means there exists a rule that matches each element in S with a unique positive integer, pairing each positive integer with exactly one element of S. (Other common terminology for this situation is that there is a "one-to-one function from S onto the positive integers" or a "bijection from S to the positive integers.") If a set is infinite but not countable, then we say that it is **uncountable**.

Examples   First we show that the set of even positive integers is countable. The required one-to-one correspondence is simply that matching 2 with 1, 4 with 2, 6 with 3, and, in general, the even integer $2n$ with n. The following table shows how the matching starts.

even positive integer	2	4	6	8	10	12	14
positive integer	1	2	3	4	5	6	7

There is something somewhat paradoxical about this matching, since it indicates that in some sense there are the "same number" of even positive integers as positive integers, in spite of the fact that the first set is a proper subset of the second. After all, two finite sets have the same number of elements if and only if they can be matched up in a one-to-one fashion.

As another example we will show that the set of all integers is countable. This time we will match 0 with 1, 1 with 2, -1 with 3, 2 with 4, -2 with 5, etc. The following table shows how the matching continues.

integer	0	1	-1	2	-2	3	-3	4	-4	5
positive integer	1	2	3	4	5	6	7	8	9	10

In this example it is not so easy to give a simple algebraic rule for the matching. An algebraic formula is not necessary, however, so long as it is clear the set is matched with the positive integers in a one-to-one fashion.

Uncountable sets also exist, but we will not give an example of one until later in this section.

Clearly any countable set is infinite, since it can be matched with the infinite set of positive integers. In some sense the smallest infinite sets are those that are countable. The following theorem makes this statement more precise.

THEOREM   Let S be any infinite set. Then S contains a countable subset.

Proof   Choose an element x_1 from S. Let S_1 be the set formed by removing x_1 from S. Now choose x_2 from S_1. Let S_2 be the set formed by removing x_2 from S_1. Choose x_3 from S_2. In general we choose x_{i+1} from S_i, and then define S_{i+1} to be S_i with x_{i+1} removed.

Notice that if we continue in this way each set S_i is infinite because it differs from S by a finite number of elements. Thus the process will never end and we can define an infinite sequence of distinct elements $x_1, x_2, \ldots$ of S this way. Let T be the set of all these elements $x_1, x_2, \ldots$. Then matching x_i with the positive integer i shows that T is countable.

Note that the proof just given does not show that S itself is countable, since there may be elements of S that never get put into T. Of course we can put any *particular* element of S into T if we want to, but this does not mean that we can exhaust S this way. If S is uncountable then it cannot equal T, which we just showed to be countable. Of course we have not exhibited a single uncountable set yet. Actually the set of real numbers is uncountable, as will follow from the next theorem. First we will introduce a notation for an idea that was used back in Section 1.2 in the proof of the Division Algorithm, namely writing a real number as the sum of an integer and a number between 0 and 1.

DEFINITIONS   integer part, fractional part

Let X be any real number. By the **integer part** of X we mean the greatest integer n such that $n \leq X$. We denote this integer n by $[X]$. For example $[3.6] = 3$, $[\pi] = 3$, $[17] = 17$, and $[-3.68] = -4$. The **fractional part** of X is defined to be $X - [X]$. Thus the fractional part of 3.6 is .6, the fractional part of -3.68 is $-3.68 - (-4) = .32$, the fractional part of 17 is 0, and the fractional part of $\sqrt{2}$ is the (irrational) number .41421356 . . . . It is easy to see that the fractional part of every real number is greater than or equal to 0 and strictly less than 1.

An Uncountable Set

It may be hard to see how one would prove a set was *not* countable, since this would mean showing that no one-to-one correspondence between the set and the positive integers existed. What is called for is a proof by contradiction.

THEOREM (Cantor)   The set of real numbers is uncountable.

Proof   Denote by R the set of real numbers. Let us suppose R is countable. Then there exists a one-to-one matching between R and the positive integers. Now some element of R is matched with 1; let us call it X_1. In general let X_i denote the element of R matched with the positive integer i. Then by the definition of a one-to-one correspondence the real numbers

$$X_1, X_2, X_3, \ldots$$

comprise all of R.

Each of these real numbers may be represented as the integer $[X_i]$ plus its fractional part where the latter is greater than or equal to 0 and less than 1. We will represent this fractional part as a decimal. Let us suppose

$$X_1 = [X_1] + .a_{11}a_{12}a_{13}\ldots,$$

$$X_2 = [X_2] + .a_{21}a_{22}a_{23}\ldots,$$

and, in general,

$$X_i = [X_i] + .a_{i1}a_{i2}a_{i3}\ldots,$$

where each a_{ij} is a digit from 0 to 9.

Now we will derive a contradiction by showing that there is an element of R that is not one of the numbers X_i. This is the number Y with the decimal representation

$$Y = .b_1b_2b_3\ldots,$$

where the digits b_i are defined as follows:

$$b_i = \begin{cases} 7 & \text{if } a_{ii} = 0, 1, 2, 3, \text{ or } 4 \\ 4 & \text{if } a_{ii} = 5, 6, 7, 8, \text{ or } 9. \end{cases}$$

The number Y is defined so as to differ from a_1 in its first digit to the right of the decimal point, from a_2 in its second digit to the right of the decimal point, etc.

Just to illustrate the definition of Y, let us suppose the first few numbers $X_1, X_2, \ldots$ are as follows.

$$X_1 = \ 17 + .⓪7\,3\,2\,9\,0\,0\ldots$$

$$X_2 = \ \ \ 3 + .5\,⑥3\,7\,4\,5\,5\ldots$$

$$X_3 = -6 + .1\,9\,⑦4\,4\,6\,2\ldots$$

$$X_4 = \ \ \ 0 + .8\,0\,0\,①7\,8\,3\ldots$$

$$X_5 = 594 + .2\,4\,7\,6\,⑦9\,6\ldots$$

The digits important for the definition of the b_i's are circled. Then $a_{11} = 0$, so $b_1 = 7$. But $a_{22} = 6$, so $b_2 = 4$. In this example we see that Y starts out

$$Y = .74474\ldots.$$

The contradiction is now that Y is in R but Y is not matched with any positive integer, contrary to our assumption. For if Y were matched with the positive integer i, then we would have $Y = X_i$. But Y differs from X_i in its ith digit, and so cannot equal X_i. This contradiction shows that R is uncountable.

Infinite sets have many properties that may at first seem strange. Putting two of them together, for example, does not necessarily yield more elements than are in just one of the sets.

THEOREM Let S and T be countable sets with no elements in common. Then the union of S and T (that is, the set of all elements in either set) is also countable.

Proof Since S is countable its elements can be matched with the positive integers in a one-to-one fashion. Let X_1 be the element of S matched with 1, X_2 be the element matched with 2, etc. Then

$$S = \{X_1, X_2, \ldots\}.$$

Likewise let T consist of the distinct elements $Y_1, Y_2, \ldots$. We match the elements of the union of S and T with the positive integers as indicated in the following table.

element	X_1	Y_1	X_2	Y_2	X_3	Y_3	X_4	Y_4	$\cdots$
positive integer	1	2	3	4	5	6	7	8	$\cdots$

This matching shows that the union of the sets is countable.

THEOREM The set of rational numbers is countable.

Proof We need to match the rational numbers with the positive integers. The following array, which lists first those rationals with denominator 1, then those with denominator 2, etc., is the key to our method.

$$\frac{0}{1}, \frac{1}{1}, \frac{-1}{1}, \frac{2}{1}, \frac{-2}{1}, \frac{3}{1}, \frac{-3}{1}, \frac{4}{1}, \ldots$$

$$\frac{0}{2}, \frac{1}{2}, \frac{-1}{2}, \frac{2}{2}, \frac{-2}{2}, \frac{3}{2}, \frac{-3}{2}, \frac{4}{2}, \ldots$$

$$\frac{0}{3}, \frac{1}{3}, \frac{-1}{3}, \frac{2}{3}, \frac{-2}{3}, \frac{3}{3}, \frac{-3}{3}, \frac{4}{3}, \ldots$$

$$\frac{0}{4}, \frac{1}{4}, \frac{-1}{4}, \frac{2}{4}, \frac{-2}{4}, \frac{3}{4}, \frac{-3}{4}, \frac{4}{4}, \ldots$$

$$\cdots$$

Of course there are repeats in this table, but these will be eliminated when we define our matching with the positive integers. Our matching will follow the pattern shown below.

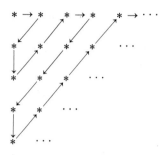

We match the rational numbers with the positive integers according to this pattern, skipping any rationals we have already used. The correspondence starts as follows.

rational	$\frac{0}{1}$,	$\frac{1}{1}$,	$\frac{1}{2}$,	$\frac{-1}{1}$,	$\frac{2}{1}$,	$\frac{-1}{2}$,	$\frac{1}{3}$,	$\frac{1}{4}$,	$\frac{-1}{3}$,	$\frac{-2}{1}$,...
integer	1,	2,	3,	4,	5,	6,	7,	8,	9,	10, ...

This one-to-one correspondence demonstrates that the rational numbers are countable.

Corollary The set of irrational numbers is uncountable.

Proof We have just seen that the rationals form a countable set. If the irrationals were also countable, then the real numbers, being the disjoint union of two countable sets, would also be countable by our theorem about such unions. But we know the real numbers are uncountable.

PROBLEMS FOR SECTION 6.1

 In the first eight problems compute the integer part and fractional part of the given number.

1. 64.716
2. 13/7
3. −4.333
4. −7/11
5. $(\sqrt{13} - 1)^2 + 2(\sqrt{13} - 1)$
6. −3 + .012
7. 4 − .01
8. $10^{1/3}$

In the next eight problems tell whether the indicated rule gives a one-to-one correspondence between the set S and the positive integers. If it does not, tell what is wrong.

9. S = the nonnegative integers; n matches with $n + 1$
10. S = the positive integers; n matches with n^2
11. S = the positive integers congruent to 2 (mod 3); $3n + 2$ matches with n
12. S = the positive integers; n matches with the least residue of n^2 (mod $n + 3$)
13. S = the integers; n matches with $2n$ if $n > 0$ and $1 - 2n$ otherwise
14. S = the integers; n matches with $2|n| + n + 1$
15. S = the odd positive integers; if n is in S, then n matches with $(n + 1)/2$
16. S = the positive integers; n matches with $[(n^2 + 1)^{1/2}]$

In the next two problems let X_1, X_2, and X_3 be as in the proof that the set of real numbers is uncountable. Find the first 3 digits of Y, where Y is as in that proof.

17. $X_1 = .12357\ldots, X_2 = .61728, X_3 = .98765\ldots$
18. $X_1 = 1/3, X_2 = \sqrt{2}, X_3 = 2^{-10}$

The next two problems refer to the proof that the rational numbers are countable.

19. What rational numbers match with the positive integers 11 and 12?
20. What rational numbers match with 13 and 14?

B

21. Show that the theorem about the union of two countable sets is valid even if the sets may have elements in common.
22. Some numbers may be represented as a decimal in more than one way. For example $.4000\ldots = .3999\ldots$ In the proof that the reals are uncountable, any decimal representation may be used for the fractional parts of the numbers X_i. Show that this ambiguity does not invalidate the proof given.

In the next four problems show the given set S is countable by defining a one-to-one correspondence between it and the positive integers.

23. S = all positive integers congruent to 3 (mod 5)
24. $S = \{1, 4, 9, 16, 25, \ldots\}$
25. $S = \{1, 2, 4, 8, 16, \ldots\}$
26. S = all integers congruent to 2 (mod 4)

27. Show that if X is any real number, then $X - 1 < [X] \le X$.

True-False. In the next eight problems tell which statements are true and give counterexamples for those that are false. Assume that $X, Y,$ and Z are real numbers and n is an integer.

28. $[X + Y] = [X] + [Y]$
29. $[X + n] = [X] + n$
30. $[-X] = -[X]$
31. $[-X] = -[X] - 1$
32. $[X] \le X < [X] + 1$

33. $[X] + [Y] \leq [X + Y] \leq [X] + [Y] + 1$
34. $[X + Y + Z] \leq [X] + [Y] + [Z] + 1$
35. If $[X] = [Y]$, then $|X - Y| < 1$.

36. Show that if A and B are real numbers such that $A < B$, then there exists a rational number X between A and B.
37. Do the previous problem with "rational" replaced by "irrational."

C 38. Suppose $S_1, S_2, \ldots, S_n$ are countable sets. Show that their union is countable.
39. Let S_i be a countable set for each positive integer i. Show that the union of all the sets S_i is countable.
40. Let S be the set of all finite subsets of the positive integers. Show that S is countable. (**Hint:** For each n there are finitely many subsets of $\{1, 2, \ldots, n\}$. Adapt the proof that the rationals are countable.)
41. Let S be the set of all subsets of the positive integers. Show that S is uncountable. (**Hint:** Associate each element T of S with a sequence of 0's and 1's, $a_1, a_2, \ldots$, where $a_i = 1$ if and only if i is in T. Adapt the proof that the reals are uncountable.)
42. Let $e = 1 + 1/1! + 1/2! + 1/3! + \cdots$. Show that if k is a positive integer, then $k!e$ is not an integer. (**Hint:** Show that $(k + 1)^{-1} + (k + 1)^{-1}(k + 2)^{-1} + (k + 1)^{-1}(k + 2)^{-1}(k + 3)^{-1} + \cdots < 1$ by comparing the expression on the left to a geometric series.)
43. Show that e is irrational. (See the previous problem.)
44. A real number is **algebraic** if it satisfies an equation $P(x) = 0$, where $P(x)$ is a nonconstant polynomial with integer coefficients. Show that the set of real numbers satisfying such an equation with the degree of P not more than n is countable for each positive integer n. (**Hint:** Problem 39.)
45. Show that the set of algebraic numbers is countable, and thus there exist **transcendental** numbers, i.e., real numbers that are not algebraic.

6.2 Finite Continued Fractions

The Euclidean Algorithm Revisited

In Section 1.3 we learned how to apply the Euclidean Algorithm to the integers a and b, $a > 0$, to find (a, b), then solve the equations backwards to find x and y such that $ax + by = (a, b)$. The process of solving the equations backwards is cumbersome and invites mistakes, so we will consider another method here.
Recall that the Euclidean Algorithm equations have the form

$$b = aq_1 + r_1,$$

$$a = r_1 q_2 + r_2,$$

$$r_1 = r_2 q_3 + r_3,$$

$$\vdots$$

$$r_{i-2} = r_{i-1} q_i + r_i,$$

$$\vdots$$

Our idea will be to use the first equation to write r_1 as a linear combination of a and b, then substitute this into the second equation to write r_2 as a linear combination of a and b, and so on, working from the top down instead of from the bottom up. For each i we seek integers x_i and y_i such that

$$r_i = ax_i + by_i.$$

For example from the first equation we have $r_1 = a(-q_1) + b$, so we take

$$x_1 = -q_1, \qquad y_1 = 1. \tag{1}$$

Likewise from the second equation we have

$$r_2 = a - r_1 q_2 = a - (-aq_1 + b)q_2$$
$$= a(1 + q_1 q_2) - bq_2,$$

so we take

$$x_2 = 1 + q_1 q_2, \qquad y_2 = -q_2. \tag{2}$$

Assume we have found x_r and y_r for $r = 1, 2, \ldots, i - 1$. Then from the equation with remainder r_i we need

$$r_i = r_{i-2} - r_{i-1} q_i$$
$$= ax_{i-2} + by_{i-2} - (ax_{i-1} + by_{i-1})q_i$$
$$= a(x_{i-2} - q_i x_{i-1}) + b(y_{i-2} - q_i y_{i-1}).$$

Thus we need to have

$$x_i = x_{i-2} - q_i x_{i-1}, \qquad y_i = y_{i-2} - q_i y_{i-1}. \tag{$*$}$$

We can use (1) and (2) to define x_i and y_i for $i = 1$ and 2, and then calculate the values for larger i from $(*)$. Since the formulas of (1) and (2) may be hard to remember, we will define "phony" values of x_i and y_i for $i = 0$ and -1, consistent with $(*)$, which will be simpler and from which the $i = 1$ and 2 values can be computed.

In particular, we want to define x_0 and y_0 so that

$$x_2 = 1 + q_1q_2 = x_0 - q_2x_1 = x_0 + q_2q_1,$$

and

$$y_2 = -q_2 = y_0 - q_2y_1 = y_0 - q_2,$$

from which we see that we can take $x_0 = 1$ and $y_0 = 0$.

In the same way we want to define x_{-1} and y_{-1} so that

$$x_1 = -q_1 = x_{-1} - q_1x_0 = x_{-1} - q_1,$$

and

$$y_1 = 1 = y_{-1} - q_1y_0 = y_{-1},$$

so we will take $x_{-1} = 0$ and $y_{-1} = 1$.

THEOREM   Let the Euclidean Algorithm be applied to a and b, $a > 0$, producing quotients q_i and remainders r_i. Define integers x_i and y_i by

$$x_{-1} = 0, \qquad y_{-1} = 1, \qquad x_0 = 1, \qquad y_0 = 0,$$

$$x_i = x_{i-2} - q_ix_{i-1}, \qquad y_i = y_{i-2} - q_iy_{i-1} \qquad \text{for} \quad i \geq 1.$$

Then $r_i = ax_i + by_i$ for $i > 0$. In particular, if r_n is the last nonzero remainder, then

$$(a, b) = ax_n + by_n.$$

Example   Let us take $a = 158$ and $b = 188$. In Section 1.3 the Euclidean Algorithm was applied to these numbers, producing quotients 1, 5, 3, 1, and 3, and remainders 30, 8, 6, 2, and 0. (The reader should review this calculation.) Thus $(158, 188) = r_4 = 2$. We will set up a table to compute the values of x_i and y_i.

i	q_i	$x_i = x_{i-2} - q_ix_{i-1}$	$y_i = y_{i-2} - q_iy_{i-1}$
-1		0	1
0		1	0
1	1	$0 - 1 \cdot 1 = -1$	$1 - 0 \cdot 1 = 1$
2	5	$1 - 5(-1) = 6$	$0 - 5 \cdot 1 = -5$
3	3	$-1 - 3 \cdot 6 = -19$	$1 - 3(-5) = 16$
4	1	$6 - 1(-19) = 25$	$-5 - 1 \cdot 16 = -21$
5	3	$-19 - 3 \cdot 25 = -94$	$16 - 3(-21) = 79$

From the $i = 4$ line we see that $158x + 188y = (158, 188) = 2$ for $x = 25$ and $y = -21$. These are the same values we got by solving the equations backwards in Section 1.3. Computing the $i = 5$ line was not necessary but does provide a check. Notice that since $r_5 = r_{n+1} = 0$ we have $ax_5 + by_5 = 158(-94) + 188 \cdot 79 = 0$. In general if r_n is the last nonzero remainder we have $ax_{n+1} + by_{n+1} = 0$, so $x_{n+1}/y_{n+1} = -b/a$.

One advantage this method has over that of solving the Euclidean Algorithm equations backwards is that once a quotient q_i is used in the calculation it is not needed any longer and can be forgotten. In fact the calculation of the x's and y's can be done as the Euclidean Algorithm equations are derived. This would be especially useful in a computer program, since using the previous method would require the computer to remember all the quotients and remainders before it could start to find x and y.

Continued Fractions

In our example the x's and y's alternated in sign, even though all the quotients q_i were positive. To simplify some later calculations we will introduce some new variables that will turn out to be positive when all the q's are positive. We define

$$h_i = (-1)^i x_i, \qquad k_i = (-1)^{i+1} y_i,$$

for $i = -1, 0, \ldots$. We can use (*) to get relations among these new variables. In particular

$$
\begin{aligned}
h_i &= (-1)^i (x_{i-2} - q_i x_{i-1}) \\
&= (-1)^2 ((-1)^{i-2} x_{i-2} + q_i (-1)^{i-1} x_{i-1}) \\
&= h_{i-2} + q_i h_{i-1};
\end{aligned}
$$

and a similar relation holds for k_i. The reader should check that from the definitions of h_i and k_i and the definitions of x_i and y_i for $i = -1$ and 0 we have $h_{-1} = 0$, $k_{-1} = 1$, $h_0 = 1$ and $k_0 = 0$. Although the q's have been (and usually will be) integers, the h's and k's are defined for any finite sequence of real numbers $q_1, \ldots, q_{n+1}$.

DEFINITION   Let $q_1, q_2, \ldots, q_{n+1}$ be real numbers. We define h_i and k_i, $i = -1$, $0, \ldots, n + 1$, by

$$h_{-1} = 0, \qquad k_{-1} = 1, \qquad h_0 = 1, \qquad k_0 = 0,$$

$$h_i = h_{i-2} + q_i h_{i-1}, \qquad k_i = k_{i-2} + q_i k_{i-1} \qquad \text{for} \quad i > 0.$$

Example   Compute h_i and k_i for $i = 1, 2,$ and 3 in terms of the q's. We have

$$h_1 = h_{-1} + q_1 h_0 = 0 + q_1(1) = q_1,$$

$$k_1 = k_{-1} + q_1 k_0 = 1 + q_1(0) = 1.$$

In the same way we compute

$$h_2 = 1 + q_1 q_2,$$

$$k_2 = q_2,$$

$$h_3 = q_1 + q_3(1 + q_1 q_2),$$

$$k_3 = 1 + q_3 q_2.$$

The ratios h_i/k_i have a curious form. For example

$$\frac{h_1}{k_1} = \frac{q_1}{1} = q_1,$$

$$\frac{h_2}{k_2} = \frac{1 + q_1 q_2}{q_2} = q_1 + \frac{1}{q_2}, \qquad \text{and}$$

$$\frac{h_3}{k_3} = \frac{q_1 + q_3 + q_1 q_2 q_3}{1 + q_2 q_3}$$

$$= \frac{q_1(1 + q_2 q_3) + q_3}{1 + q_2 q_3} = q_1 + \frac{q_3}{1 + q_2 q_3} = q_1 + \cfrac{1}{q_2 + \cfrac{1}{q_3}}.$$

Since the notation is getting messy we will introduce a more compact form.

DEFINITIONS   finite continued fraction, simple finite continued fraction
By a **finite continued fraction** we mean an expression of the form

$$q_1 + \cfrac{1}{q_2 + \cfrac{1}{q_3 + \cfrac{1}{\ddots + \cfrac{1}{q_r}}}},$$

where the q's are real numbers, positive for $i > 1$. We denote this by $[q_1, q_2, \ldots, q_r]$. If the q's are all integers we say the continued fraction is **simple**.

Examples From the definition

$$[-2, 3, 5] = -2 + \cfrac{1}{3 + \cfrac{1}{5}}$$

$$= -2 + \cfrac{1}{\left(\cfrac{16}{5}\right)} = -2 + \frac{5}{16} = \frac{-27}{16}.$$

Likewise $[1, 5, 3, 1, 3] =$

$$1 + \cfrac{1}{5 + \cfrac{1}{3 + \cfrac{1}{1 + \cfrac{1}{3}}}} = 1 + \cfrac{1}{5 + \cfrac{1}{3 + \cfrac{3}{4}}} = 1 + \cfrac{1}{5 + \cfrac{4}{15}} = 1 + \frac{15}{79} = \frac{94}{79}.$$

Usually we will be interested in only simple continued fractions, as in our examples. Requiring the $q_i > 0$ for $i > 1$ means that we do not have to worry about dividing by zero in our expression. Notice that the definition implies that

$$[q_1, q_2, \ldots, q_r] = \left[q_1, q_2, \ldots, q_{r-1} + \frac{1}{q_r}\right].$$ (**)

Now we have a compact way of expressing the pattern we discovered for the ratios h_i/k_i, namely

$$\frac{h_1}{k_1} = [q_1],$$

$$\frac{h_2}{k_2} = [q_1, q_2], \quad \text{and}$$

$$\frac{h_3}{k_3} = [q_1, q_2, q_3].$$

THEOREM 6.1 Let $q_1, q_2, \ldots, q_r$ be real numbers, with $q_i > 0$ for $i > 1$, and let h_i and k_i be defined as above. Then

$$[q_1, q_2, \ldots, q_r] = \frac{h_r}{k_r}.$$

Proof The proof will be by induction on r. We already know the theorem is true for $r = 1$, 2, and 3. Suppose we know the theorem to be true for $r = n$.

Let $q_1, q_2, \ldots, q_{n+1}$ be given, and let h_i and k_i be defined as usual for $i = -1$, $0, \ldots, n + 1$.

We define another sequence of n real numbers $Q_1, Q_2, \ldots, Q_n$ by

$$Q_1 = q_1, \quad Q_2 = q_2, \quad \cdots, \quad Q_{n-1} = q_{n-1}, \quad \text{and} \quad Q_n = q_n + \frac{1}{q_{n+1}}.$$

Let numbers H_i and K_i be defined for $i = -1, 0, \ldots, n$ just as the h's and k's are, but in terms of the Q's instead of the q's. Note that $H_i = h_i$ and $K_i = k_i$ for $i < n$ since $Q_i = q_i$ for $i < n$. Then

$$[q_1, q_2, \ldots, q_{n+1}] = \left[q_1, \ldots, q_{n-1}, q_n + \frac{1}{q_{n+1}} \right] \quad \text{(by (**))}$$

$$= [Q_1, \ldots, Q_n] = \frac{H_n}{K_n} \quad \text{(by the induction hypothesis)}$$

$$= \frac{H_{n-2} + Q_n H_{n-1}}{K_{n-2} + Q_n H_{n-1}} = \frac{h_{n-2} + \left(\dfrac{q_{n+1}}{q_{n+1}} \right) h_{n-1}}{k_{n-2} + \left(\dfrac{q_{n+1}}{q_{n+1}} \right) k_{n-1}}$$

$$= \frac{h_{n-2} + q_n h_{n-1} + \dfrac{h_{n-1}}{q_{n+1}}}{k_{n-2} + q_n k_{n-1} + \dfrac{k_{n-1}}{q_{n+1}}}$$

$$= \frac{h_n + \dfrac{h_{n-1}}{q_{n+1}}}{k_n + \dfrac{k_{n-1}}{q_{n+1}}} = \frac{h_{n-1} + q_{n+1} h_n}{k_{n-1} + q_{n+1} k_n} = \frac{h_{n+1}}{k_{n+1}}.$$

Example Compute $[-2, 3, 5]$ by means of Theorem 6.1.

We have $[-2, 3, 5] = h_3/k_3$, where the computation of the h's and k's is displayed in the following table.

i	q_i	$h_i = h_{i-2} + q_i h_{i-1}$	$k_i = k_{i-2} + q_i k_{i-1}$
-1		0	1
0		1	0
1	-2	$0 + (-2)1 = -2$	$1 + (-2)0 = 1$
2	3	$1 + 3(-2) = -5$	$0 + 3(1) = 3$
3	5	$-2 + 5(-5) = -27$	$1 + 5(3) = 16$

We see that $[-2, 3, 5] = -27/16$.

THEOREM 6.2   Any simple finite continued fraction is equal to a rational number. Conversely, any nonintegral rational number can be expressed as a simple finite continued fraction $[q_1, q_2, \ldots, q_{n+1}]$, where $q_{n+1} > 1$.

Proof   The first statement follows from Theorem 6.1 since if $q_1, \ldots, q_r$ are all integers, then so are all the h's and k's, and so h_r/k_r is rational. (It is also obvious from the definition of simple finite continued fraction.)

Conversely, any nonintegral rational number can be written as b/a with $a > 1$ and $(a, b) = 1$. Apply the Euclidean Algorithm as usual, getting quotients q_i and remainders r_i. Let r_n be the last nonzero remainder. Then by the first theorem of this section

$$0 = r_{n+1} = ax_{n+1} + by_{n+1},$$

so

$$\frac{b}{a} = \frac{-x_{n+1}}{y_{n+1}} = \frac{h_{n+1}}{k_{n+1}}.$$

This is $[q_1, \ldots, q_{n+1}]$ by Theorem 6.1.

The last equation in the Euclidean Algorithm is

$$r_{n-1} = q_{n+1}r_n + 0,$$

where if $n = 1$ we interpret r_0 to be a. Since $1 = (a, b) = r_n < r_{n-1}$, this equation implies that $q_{n+1} > 1$.

PROBLEMS FOR SECTION 6.2

 In the first four problems assume the given triple is q_1, q_2, q_3. _Find_ x_i _and_ y_i _for_ $i = 1, 2, 3$.

  1. 6, 1, 2   2. 4, 9, 3   3. $-2, 5, 7$ 4. $-3, 1, 10$

In the next six problems find x _and_ y _such that_ $ax + by = (a, b)$ _by the method of the first theorem of this section._

  5. $a = 51, b = 71$ 6. $a = 92, b = 300$
  7. $a = 731, b = 45$ 8. $a = 666, b = 101$
  9. $a = 402, b = -1005$ 10. $a = 671, b = -616$

In the next four problems assume the given triple is q_1, q_2, q_3. _Find_ h_i _and_ k_i _for_ $i = 1, 2, 3$.

  11. 2, 3, 4   12. 5, 4, 3   13. $-4, 1, 6$ 14. 0, 5, 7

*In the next four problems evaluate the given continued fraction as a rational num-
ber by the method of the examples after the definition of a finite continued fraction.
Then confirm your work by using Theorem 6.1.*

15. $[2, 4, 1, 5]$ 16. $[6, 1, 1]$
17. $[0, 3, 7, 5, 2]$ 18. $[-5, 2, 1, 2]$

*In the next six problems write each rational number b/a as a continued fraction
$[q_1, \ldots, q_{n+1}]$ by applying the Euclidean Algorithm to a and b.*

19. $73/41$ 20. $41/73$ 21. $-45/14$
22. $-34/105$ 23. 3.14 24. 3.1416

B *Define $R_i = h_i/k_i$. In each of the next four problems let the number given be b/a.
Let $q_1, \ldots, q_{n+1}$ be the quotients when the Euclidean Algorithm is applied to a and
b. Find R_i for $i = 1$ to $n + 1$, and label these points on the real axis, along with b/a.*

25. $71/15$ 26. $101/43$ 27. 3.14 28. 2.72

29. Find expressions for h_4 and k_4 in terms of the q's.
30. Show that if $q_r > 1$, then $[q_1, \ldots, q_r] = [q_1, \ldots, q_{r-1}, q_r - 1, 1]$.
31. Prove that k_i does not depend on q_1.
32. Prove that if $[q_1, \ldots, q_r]$ is a finite continued fraction, then $k_i > 0$ for $i = 1, 2, \ldots, r$.
33. Prove that $h_i = (1 + q_i q_{i-1})h_{i-2} + q_i h_{i-3}$ for $i > 1$.
34. Show that $k_i > k_{i-1}$ for $i > 2$ if $q_i > 0$ for $i > 1$.

C *If the rational number b/a is expressed as the simple continued fraction
$[q_1, \ldots, q_{n+1}]$ with $q_{n+1} > 0$, then it also has the representation
$[q_1, \ldots, q_{n+1} - 1, 1]$, as is shown in a previous problem. It turns out that every
rational number has exactly these two representations, and no more. This is proved
in the next four problems.*

35. Show that each rational number has at least two distinct representations
    as a simple finite continued fraction.
36. Suppose $R = [a_1, \ldots, a_m] = [b_1, \ldots, b_n]$, where these are simple finite con-
    tinued fractions with $m > 1$, $n > 1$, $a_m > 1$, and $b_n > 1$. Show that $a_1 = b_1$
    and $[a_2, \ldots, a_m] = [b_2, \ldots, b_n]$. (**Hint:** Show $a_1 = [R]$, where this is the
    greatest integer function defined in Section 6.1.)
37. Show that under the hypothesis of the last problem $m = n$ and $a_i = b_i$ for
    $i = 1, \ldots, m$.
38. Show that any rational number can be expressed as a finite simple continued
    fraction in exactly two ways.

6.3 Infinite Continued Fractions

In the last section we saw that every finite sequence of integers $q_1, \ldots, q_r$,
with $q_i > 0$ for $i > 1$, corresponds to a rational number expressed by the continued fraction $[q_1, \ldots, q_r]$ and that each simple finite continued fraction rep-

resents a rational number. In this section we will try to make sense out of an expression of the form $[q_1, q_2, \ldots]$, where the q's form an infinite sequence. To understand this section fully the reader should be familiar with the meaning of a statement of the form

$$\lim_{n \to \infty} R_n = L,$$

where $R_1, R_2, \ldots$ is an infinite sequence of real numbers, but a useful, if somewhat more vague, understanding of infinite continued fractions is possible even without a knowledge of limits.

It turns out that there is a one-to-one correspondence between infinite continued fractions and the irrational numbers, so our analysis will consist of two parts: (1) showing how each infinite sequence of q's defines an irrational number, and (2) showing how each irrational number produces an infinite sequence of q's.

Making Sense out of $[q_1, q_2, \ldots]$

A natural definition for $[q_1, q_2, \ldots]$ would be as the limit of $[q_1, \ldots, q_n]$ as $n \to \infty$, but, of course, it is not obvious whether the limit exists or not. The key to our analysis will be the result from the previous section that $[q_1, \ldots, q_n] = h_n/k_n$, where the h's and k's are as defined there. We need to know more about the relations among these variables.

Since some of our results will apply to finite continued fractions as well as infinite continued fractions, we will assume for the time being that $q_1, q_2, \ldots$ is a finite or infinite sequence of integers, with $q_i > 0$ for $i > 1$ (whenever q_i is defined, of course).

LEMMA 1   We have $1 = k_1 \leq k_2 < k_3 < \cdots$.

Proof   In Section 6.1 we computed that $k_1 = 1$ and $k_2 = q_2$, which is at least 1 by our assumption about the q's. The equation $k_i = k_{i-2} + q_i k_{i-1}$ then shows that $k_i \geq k_{i-2} + k_{i-1} > k_{i-1}$ for $i > 2$.

Consider the equations

$$h_i = h_{i-2} + q_i h_{i-1},$$

$$k_i = k_{i-2} + q_i k_{i-1}.$$

Suppose we eliminate q_i by multiplying the first equation by k_{i-1}, the second by h_{i-1}, and subtracting. We get

$$k_{i-1} h_i - h_{i-1} k_i = k_{i-1} h_{i-2} - h_{i-1} k_{i-2} = -(k_{i-2} h_{i-1} - h_{i-2} k_{i-1}),$$

which is just the negative of what we started with, but with i decreased by 1. Since $k_{-1} h_0 - h_{-1} k_0 = 1 \cdot 1 - 0 \cdot 0 = 1$, we have $k_0 h_1 - h_0 k_1 = -1$, $k_1 h_2 - h_1 k_2 = 1$, etc.

LEMMA 2   We have $k_{i-1}h_i - h_{i-1}k_i = (-1)^i$. In consequence, $(h_i, k_i) = 1$.

The second statement of this lemma, which follows from Theorem 1.4, says that the fraction h_i/k_i is in lowest terms. We will denote this fraction by R_i. By Lemma 1, R_i makes sense for $i > 0$. Dividing the equation of Lemma 2 by $k_i k_{i-1}$ yields

$$R_i - R_{i-1} = \frac{(-1)^i}{k_i k_{i-1}}. \tag{$*$}$$

Likewise

$$R_{i+1} - R_i = \frac{(-1)^{i+1}}{k_{i+1}k_i},$$

and adding this to $(*)$ gives us

$$R_{i+1} - R_{i-1} = \frac{(-1)^i\left(\dfrac{1}{k_{i-1}} - \dfrac{1}{k_{i+1}}\right)}{k_i}.$$

By Lemma 1 we have

$$k_{i+1} > k_{i-1} \qquad \text{for} \quad i > 1,$$

so $R_{i+1} - R_{i-1}$ is positive or negative according as i is even or odd. Letting $i = 2, 3, \ldots$ gives $R_3 - R_1 > 0$, $R_4 - R_2 < 0$, $R_5 - R_3 > 0$, etc. This proves the following result.

LEMMA 3   We have

$$R_1 < R_3 < R_5 < \cdots \text{ and } R_2 > R_4 > R_6 > \cdots.$$

Example   Consider the finite continued fraction $[1, 2, 3, 1]$. Compute the R's and graph them on the real axis.

The h's, k's, and R's are computed in the following table.

i	-1	0	1	2	3	4
q_i			1	2	3	1
h_i	0	1	1	3	10	13
k_i	1	0	0	2	7	9
R_i			1	$3/2 = 1.5$	$10/7 \approx 1.43$	$13/9 \approx 1.44$

LEMMA 4 Suppose i and j are positive integers, with i even and j odd. Then $R_i > R_j$.

Proof Let n be the larger of i and j. From (∗) and Lemma 1 the quantity $R_n - R_{n-1}$ is positive or negative according as n is i or j. Then by Lemma 3 we have in the first case

$$R_i > R_{i-1} \ge R_j,$$

and in the second case

$$R_j < R_{j-1} \le R_i.$$

THEOREM 6.3 Let $q_1, q_2, \ldots$ be an infinite sequence of integers, with $q_i > 0$ for $i > 1$. Then

$$\lim_{r \to \infty} [q_1, q_2, \ldots, q_r]$$

exists, and is an irrational number.

Proof By Lemmas 3 and 4 the numbers $R_1, R_3, \ldots$ form an increasing sequence of real numbers that is bounded above by R_2, and so has a limit. Likewise the $R_2, R_4, \ldots$ is a decreasing sequence bounded below by R_1, and so also has a limit. Since Lemma 1 implies the k's go to infinity, by (∗) these two limits must be the same, say the real number R. This proves the first statement of the theorem. Note that

$$R_1 < R_3 < \cdots \le R \le \cdots < R_4 < R_2, \qquad (**)$$

so R is not equal to any quotient R_i.

Now we prove that R is irrational. Suppose $R = b/a$, with $a > 0$. From (∗∗) b/a is strictly between R_i and R_{i-1}, and so

$$0 < \left| \frac{b}{a} - R_{i-1} \right| < |R_i - R_{i-1}|.$$

The quantity on the right equals $1/k_i k_{i-1}$ by (∗). If we multiply through by ak_{i-1} we get

$$0 < |bk_{i-1} - ah_{i-1}| < \frac{a}{k_i},$$

where we have used the fact that $R_{i-1} = h_{i-1}/k_{i-1}$.

Since we can make k_i as large as we please by taking i large, we can make the quantity on the right less than 1. But the quantity inside the absolute value signs is an integer. We have an integer strictly between 0 and 1, which is impossible. This contradiction shows that R must be irrational.

DEFINITIONS   infinite continued fraction, convergent

Let $q_1, q_2, \ldots$ be an infinite sequence of integers, positive except perhaps for q_1. We define

$$[q_1, q_2, \ldots] = \lim_{n \to \infty} [q_1, \ldots, q_n].$$

This is called an **infinite continued fraction**. The numbers $[q_1, q_2, \ldots, q_n]$, $n = 1, 2, \ldots$, are said to be the **convergents** of the infinite continued fraction.

Given a sequence of q's, it is possible to approximate the corresponding irrational number $[q_1, q_2, \ldots]$ as closely as one wants by computing the numbers R_i. In fact by (*) and (**) the error in using R_{i-1} to approximate R is less than $1/k_i k_{i-1}$. Usually it is not possible to name R other than by the continued fraction, however. An exception occurs when the sequence of q's repeats.

Example   What irrational number does $[1, 2, 1, 2, 1, 2, \ldots]$ represent?
Call the number R. Note that

$$R_{2n} = 1 + \cfrac{1}{2 + \cfrac{1}{R_{2n-2}}}.$$

(Write out $R_2 = [1, 2]$, $R_4 = [1, 2, 1, 2]$, and $R_6 = [1, 2, 1, 2, 1, 2]$ to get the idea.) If we let n go to infinity we see that

$$R = 1 + \cfrac{1}{2 + \cfrac{1}{R}}.$$

This leads to the equation

$$2R^2 - 2R - 1 = 0,$$

which has the solutions $R = (1 \pm \sqrt{3})/2$. Since R is positive taking the plus sign gives its value.

Finding the Continued Fraction for an Irrational Number

Now we will consider the converse problem of finding the infinite continued fraction corresponding to a given irrational number. Let the number be S. We need a sequence of q's such that $S = [q_1, q_2, \ldots]$.

If S were rational, say b/a, we would find the corresponding q's by applying the Euclidean Algorithm to a and b, getting relations

$$b = aq_1 + r_1, \quad 0 < r_1 < a,$$

$$a = r_1q_2 + r_2, \quad 0 < r_2 < r_1,$$

and so on. If we divide the first equation and inequality by a we get

$$\frac{b}{a} = q_1 + \frac{r_1}{a}, \quad 0 < \frac{r_1}{a} < 1.$$

We see that q_1 is the integer part and r_1/a the fractional part of b/a, that is

$$q_1 = \left[\frac{b}{a}\right], \quad \frac{r_1}{a} = \frac{b}{a} - \left[\frac{b}{a}\right].$$

In the same way dividing the second equation and inequality by r_1 gives

$$\frac{a}{r_1} = q_2 + \frac{r_2}{r_1}, \quad 0 < \frac{r_2}{r_1} < 1.$$

Thus $q_2 = [a/r_1] = [1/(r_1/a)] = [1/(b/a - [b/a])]$.

We need some notation to simplify what is going on here. Let $S_1 = a/b$. Then

$$q_1 = [S_1].$$

Also if we define S_2 to be $1/(S_1 - q_1)$, then

$$q_2 = [S_2].$$

The generalization of this process gives us a method for finding the continued fraction expression for an irrational number.

THEOREM 6.4   Let S be a real number. Let $S_1 = S$ and $q_1 = [S_1]$. Suppose S_i and q_i have been defined for $i = 1, 2, \ldots, r$. If S_i does not equal q_i, then define S_{i+1} and q_{i+1} by

$$S_{i+1} = \frac{1}{S_i - q_i}, \quad q_{i+1} = [S_{i+1}].$$

Then if S is rational this process stops with the definition of some q_r, and we have

$$S = [q_1, \ldots, q_r],$$

while if S is irrational an infinite sequence of q's is defined and

$$S = [q_1, q_2, \ldots].$$

Before proving this theorem we will do an example of the computation of the numbers S_i and q_i. We will take $S = \pi$, which is irrational (although this fact will not be proved in this book). Then $S_1 = S = 3.1415926535897\ldots$, and so $q_1 = [3.14\ldots] = 3$. (The first 14 digits of π may be remembered from the numbers of letters in the words "how I want a drink, alcoholic of course, after the heavy integral theoretic lecture.") Then $S_2 = 1/(\pi - 3) = 7.0625\ldots$ and $q_2 = 7$. Likewise $S_3 = 1/.0625\ldots = 15.9965\ldots$ and so $q_3 = 15$. So far we have $\pi = [3, 7, 15, \ldots]$.

In practice if we start with a value of π from a calculator or table of constants, which is only an approximation to the true value, the numbers q_i we compute will eventually diverge from the true values, how soon depending on the accuracy of the original approximation. Using a calculator that gives π to 9 digits to the right of the decimal point, for example, we compute the continued fraction $[3, 7, 15, 1, 293, 10, 3, 8, 2, 1, \ldots]$ for π. The true value, however, is known to start out $[3, 7, 15, 1, 292, 1, 1, 1, 2, 1, \ldots]$. (If there is any pattern in this expansion, no one has ever figured it out.)

The values of R_n turn out to be good rational approximations to the irrational number S. What "good" means will not be made precise here; but roughly we want R_n close to S without having too large a denominator k_n.

In this example we have $R_1 = [3] = 3$, a number claimed to be π by Leeds K. Field in his 1953 book *Mathematics, Minus and Plus*. Likewise $R_2 = [3, 7] = 22/7$, which is the usual rational approximation, and correct within about 4/100 of 1%. Another amazingly good approximation to π is $R_4 = [3, 7, 15, 1] = 355/113$, with an error of less than one part in 10 million.

Proof of Theorem 6.4   If S is rational it is not hard to see that the q's we get are exactly the q's of the Euclidean Algorithm, so we will concentrate on the case when S is irrational.

Since $S = S_1$ is irrational it cannot equal the integer q_1, and so $S_2 = 1/(S_1 - q_1)$ is defined. In fact S_2 must be irrational also, since otherwise $S_1 = q_1 + 1/S_2$ would be rational. A continuation of this argument shows that all the numbers S_i are defined and irrational, so that the sequence of q's is infinite.

Note that since $S_i - q_i$ is the fractional part of S_i, $0 < S_i - q_i < 1$, and so $S_{i+1} = 1/(S_i - q_i) > 1$. Thus $q_{i+1} = [S_{i+1}] \geq 1$ for $i > 0$.

Then by Theorem 6.3 we have $[q_1, q_2, \ldots] = R$ for some irrational number R. Let the numbers h_i, k_i, and R_i be defined as before, so that R is the limit of the numbers R_i. Note that

$$S_2 = \frac{1}{S_1 - q_1}, \quad \text{so} \quad S = S_1 = q_1 + \frac{1}{S_2},$$

and, in general,

$$S_i = q_i + \frac{1}{S_{i+1}}.$$

Then

$$S = S_1 = q_1 + \frac{1}{S_2} = q_1 + \cfrac{1}{q_2 + \cfrac{1}{S_3}} = \cdots = [q_1, \ldots, q_{n-1}, S_n].$$

By Theorem 6.1 the last expression equals

$$\frac{h_{n-2} + S_n h_{n-1}}{k_{n-2} + S_n k_{n-1}}.$$

Thus

$$\begin{aligned}
S - R_{n-1} &= \frac{h_{n-2} + S_n h_{n-1}}{k_{n-2} + S_n k_{n-1}} - \frac{h_{n-1}}{k_{n-1}} \\
&= \frac{k_{n-1}(h_{n-2} + S_n h_{n-1}) - (k_{n-2} + S_n k_{n-1})h_{n-1}}{k_{n-1}(k_{n-2} + S_n k_{n-1})} \\
&= \frac{-(k_{n-2}h_{n-1} - h_{n-2}k_{n-1})}{k_{n-1}(k_{n-2} + S_n k_{n-1})} = \frac{(-1)^n}{k_{n-1}(k_{n-2} + S_n k_{n-1})}
\end{aligned}$$

by Lemma 2. Now the k's go to infinity and $S_n > 0$ for $n > 0$ so the last quantity goes to 0 as n goes to infinity. We see

$$0 = \lim_{n \to \infty} S - R_{n-1} = S - R.$$

Expansions of Quadratic Irrationalities

If the irrational number S is of the form $A\sqrt{d} + B$, where A and B are rational numbers and d is a positive integer, then we can find its (infinite) continued fraction expansion explicitly. In fact, although we will not prove it here, it turns out that such numbers are exactly those having expansions that repeat past some point.

DEFINITION periodic continued fraction
An infinite continued fraction $[q_1, q_2, \ldots]$ is said to be **periodic** if it repeats from some point on, that is, if there exist positive integers m and r such that $q_n = q_{n+r}$ for $n > m$. An example of a periodic expansion is

$$[-4, 5, 1, 2, 7, 1, 3, 7, 1, 3, 7, 1, 3, \ldots],$$

where the pattern 7-1-3 keeps repeating. (Here $m = 4$ and $r = 3$.) We will denote this continued fraction by

$$[-4, 5, 1, 2, \overline{7, 1, 3}],$$

where the bar indicates the group of q's under it repeats indefinitely. As another example of this notation, from the example after Theorem 6.3 we have $(1 + \sqrt{3})/2 = [1, 2, \overline{1, 2}] = [1, \overline{2, 1}] = [\overline{1, 2}]$.

Example Find the continued fraction expansion of $S = (1/6)\sqrt{21} + 7/2$.

Let $S_1 = S = (\sqrt{21} + 21)/6$. This number is about 4.3, so $q_1 = [S_1] = 4$. Then

$$
\begin{aligned}
S_2 &= \left(\frac{\sqrt{21} + 21}{6} - 4\right)^{-1} = \left(\frac{\sqrt{21} - 3}{6}\right)^{-1} \\
&= \frac{6}{\sqrt{21} - 3} = \frac{6(\sqrt{21} + 3)}{21 - 9} = \frac{\sqrt{21} + 3}{2}.
\end{aligned}
$$

This number is about 3.8, so $q_2 = 3$. Then

$$
\begin{aligned}
S_3 &= \left(\frac{\sqrt{21} + 3}{2} - 3\right)^{-1} = \left(\frac{\sqrt{21} - 3}{2}\right)^{-1} \\
&= \frac{2}{\sqrt{21} - 3} = \frac{2(\sqrt{21} + 3)}{21 - 9} = \frac{\sqrt{21} + 3}{6}.
\end{aligned}
$$

Since S_3 is about 1.6 we have $q_3 = 1$. Then

$$
S_4 = \left(\frac{\sqrt{21} + 3}{6} - 1\right)^{-1} = \left(\frac{\sqrt{21} - 3}{6}\right)^{-1}.
$$

But this is the same as S_2. Thus $q_4 = q_2 = 3$, $q_5 = q_3 = 1$, and the expansion repeats from now on. We see

$$
\frac{\sqrt{21} + 21}{6} = [4, 3, 1, 3, 1, \ldots] = [4, \overline{3, 1}].
$$

Example Find the irrational number represented by $S = [-1, 1, \overline{1, 2, 3}]$.

We have

$$
S_1 = -1 + \cfrac{1}{1 + \cfrac{1}{S_3}}, \qquad \text{where} \quad S_3 = [\overline{1, 2, 3}] = 1 + \cfrac{1}{2 + \cfrac{1}{3 + \cfrac{1}{S_3}}}.
$$

Simplifying the last equation leads to

$$7S_3^2 - 8S_3 - 3 = 0,$$

which has the solutions $(4 \pm \sqrt{37})/7$. Since $S_3 > 0$ we choose the plus sign.

Finally by plugging this value of S_3 into the first equation of this example, simplifying, and rationalizing the denominator, we get

$$S_1 = S = \frac{\sqrt{37} - 11}{12}.$$

PROBLEMS FOR SECTION 6.3

 In the first eight problems find the irrational number corresponding to the given periodic continued fraction.

1. $[\overline{1}]$
2. $[\overline{2}]$
3. $[\overline{2, 1}]$
4. $[\overline{1, 3}]$
5. $[1, \overline{3}]$
6. $[-2, \overline{1}]$
7. $[-1, 3, \overline{2, 3}]$
8. $[-3, 1, \overline{1, 4}]$

In the next eight problems find the periodic continued fraction for the given irrational number.

9. $(\sqrt{5} - 1)/2$
10. $\sqrt{2}$
11. $(5 - \sqrt{13})/6$
12. $(17 - \sqrt{3})/13$
13. $2\sqrt{3}$
14. $1 - 2/\sqrt{3}$
15. $(5 + \sqrt{37})/4$
16. $(\sqrt{37} + 5)/3$

In the next four problems find the first four terms of the continued fraction expansion of the given number.

17. $\pi/4$
18. $2^{1/3}$
19. $\sin 10°$
20. e

 21. Show that if m is a positive integer, then $[\overline{m}] = (m + \sqrt{m^2 + 4})/2$.
22. Show that if a and b are positive integers, then $[\overline{a, b}] = (ab + (a^2b^2 + 4ab)^{1/2})/2b$.
23. Let A, B, A', and B' be rational numbers and d a positive integer, not a perfect square. Show that if $A\sqrt{d} + B = A'\sqrt{d} + B'$, then $A = A'$ and $B = B'$.
24. Let R be irrational. Show that R has the form $A\sqrt{d} + B$, where A and B are rational and d is a positive integer if and only if R is a solution of an equation $ax^2 + bx + c = 0$ with a, b, and c integers.
25. Show that the continued fraction expansion of an irrational number is unique.
26. Let m be a positive integer. Show that $(m^2 + 1)^{1/2} = [m, \overline{2m}]$.
27. From the proof of Theorem 6.3 we see that if $[q_1, q_2, \ldots] = R$, then $|R - R_{i-1}| < 1/k_i k_{i-1}$. Assuming $\pi = [3, 7, 15, \ldots]$, how close does this say π is to 22/7?

28. Assuming $\pi = [3, 7, 15, 1, 292, \ldots]$, how close must it be to $355/113$? (See the last problem.)

29. Let $q_i = 1$ for all $i > 0$. Show that $f_i = k_i = h_{i-1}$ for $i > 0$, where f_i is the ith Fibonacci number.

C *In the next three problems assume that $R = [\overline{q_1, \ldots, q_n}]$, where the q's are positive integers.*

30. Show that $R = [q_1, \ldots, q_n, R]$.

31. Show that if the h's and k's are defined as usual, then $R = (h_{n-2} + Rh_{n-1})/(k_{n-2} + Rk_{n-1})$, where $h_{n-2}, k_{n-2}, h_{n-1}$, and k_{n-1} are integers.

32. Show that R satisfies a quadratic equation with integer coefficients.

6.4 Decimal Representation

In our study of the continued fraction expansions of real numbers, we found a tidy matching of finite expansions with rational and infinite expansions with irrational numbers. A more familiar way of distinguishing rational from irrational numbers has to do with whether their decimal expansions repeat or not.

DEFINITIONS   periodic decimal, period, terminating decimal
Consider the decimal representation

$$\pm e_1 e_2 \cdots e_m.d_1 d_2 \cdots .$$

We say that this expansion is **periodic** if there exist integers k and $j, j > 0$, such that $d_i = d_{i+j}$ for $i > k$. If j is the least such positive integer, then we call j the **period** of the expansion, and denote the expansion by

$$\pm e_1 \cdots e_m.d_1 \cdots d_k\overline{d_{k+1} \cdots d_{k+j}}.$$

For example $2/3 = .6666 \ldots = .\overline{6}$, and the period of this expansion is 1. Likewise $.165214214214 \ldots = .165\overline{214}$, with period 3. If $j = 1$ and $d_i = 0$ for $i > k$, then we say the expansion is **terminating**. An example of a terminating expansion is $9/4 = 2.25 = 2.25\overline{0}$.

Since any real number can be written in the form $\pm(n + X)$, where n is a nonnegative integer and $0 \le X < 1$, we will concentrate on the expansions of numbers X in this range. We will assume certain familiar facts:

1. Each real number $X, 0 \le X < 1$, has a decimal expansion.

2. This expansion is unique except that the terminating decimal $.d_1 \cdots d_k\overline{0}$, with $d_k > 0$, also has the expansion $.d_1 \cdots d_{k-1}(d_k - 1)\overline{9}$. For example $.135 = .135\overline{0} = .134\overline{9} = .134999 \ldots$.

3. A decimal expansion for the rational number b/a can be found by dividing b by a using the usual long division algorithm taught in elementary school.

4. Decimals may be added and subtracted in the usual way, and multiplied or divided by powers of 10 by moving the decimal point appropriately.

THEOREM   A real number is rational if and only if it has a periodic decimal expansion.

Proof   If Y is any real number we can write $Y = \pm(n + X)$, where $0 \le X < 1$, and n is a nonnegative integer. Since Y has a periodic expansion if and only if X does, and is rational if and only if X is, it suffices to prove the theorem for numbers between 0 and 1.

We will start by assuming that X has the periodic representation

$$.d_1 d_2 \cdots d_k \overline{d_{k+1} \cdots d_{k+j}}.$$

Then $10^{k+j}X - 10^k X$

$$= d_1 \cdots d_{k+j}.\overline{d_{k+1} \cdots d_{k+j}} - d_1 \cdots d_k.\overline{d_{k+1} \cdots d_{k+j}},$$

which is an integer, say s. Then X is the rational number $s/(10^{k+j} - 10^k)$.

Conversely let X be the rational number b/a, where $0 \le b < a$ and $(a, b) = 1$. We can find a decimal expansion for X by the usual algorithm, which we illustrate below for $b = 7$ and $a = 22$.

```
           0.3 1 8 1 8 1 8 ···
      22 ) 7.0 0 0 0 0 0 0 ···
           6 6
            ④0
            2 2
            ①⑧0
            1 7 6
              ④0
              2 2
              ①⑧0
               ···
```

The circled numbers are the remainders at each stage of the algorithm, and each is nonnegative and less than the divisor, 22. In general each remainder r satisfies $0 \le r < a$. Since there are only finitely-many possible remainders, eventually one must repeat, and from that point on the digits of the dividend will also repeat, producing a periodic expansion. In the example the remainder 4 repeats first, giving the periodic expansion $7/22 = .3\overline{18}$.

Example   Express $X = .37\overline{216}$ as the quotient of two integers.
We follow the proof of the above theorem with $k = 2$ and $j = 3$, noting that

$$10^5 X - 10^2 X = 37216.\overline{216} - 37.\overline{216} = 37179.$$

Then $X = 37179/(10^5 - 10^2) = 37179/99900 = 1377/3700$. This can be checked by dividing out the latter fraction.

The Period of a Rational Number

The long division algorithm illustrated in the proof of the last theorem deserves closer study. If the divisor is a, at each stage of the algorithm we get a remainder r_i satisfying $0 \leq r_i < a$. Then r_i is multiplied by 10, and $10r_i$ is divided by a to get a new remainder r_{i+1}. We have

$$10r_i = q_{i+1}a + r_{i+1}, \quad 0 \leq r_{i+1} < a,$$

where q_{i+1} is the next digit of the quotient. Continuing the process, the next quotient digit q_{i+2} and remainder r_{i+2} satisfy

$$10r_{i+1} = q_{i+2}a + r_{i+2}, \quad 0 \leq r_{i+2} < a.$$

Notice that from these equations

$$r_{i+1} \equiv 10r_i \pmod{a} \quad \text{and} \quad r_{i+2} \equiv 10r_{i+1} \pmod{a}.$$

Substituting the first congruence into the second yields

$$r_{i+2} \equiv 10^2 r_i \pmod{a}.$$

The reader should check that in the same way if r_{i+3} is the next remainder, then

$$r_{i+3} \equiv 10^3 r_i \pmod{a}.$$

In general, if $r_1, r_2, \ldots$ are the remainders produced when b is divided by a using the long-division algorithm, where $0 \leq b < a$ and $(a, b) = 1$, then

$$r_{i+j} \equiv 10^j r_i \pmod{a}$$

for all positive integers i and j.

In particular, when a remainder repeats, say when $r_{i+j} = r_i$, we have

$$r_i \equiv 10^j r_i \pmod{a}.$$

It would be nice to cancel the r_i from both sides of this congruence, but in order to do this we need to know that $(r_i, a) = 1$.

Let us see how the remainders r_i start out. Since we are assuming that $b < a$, the first step in the long-division algorithm is to divide $10b$ by a, getting a quotient digit q_1 (which may be 0) and remainder r_1 satisfying

$$10b = q_1 a + r_1.$$

A good start would be to show that $(r_1, a) = 1$. Suppose some prime p divides both r_1 and a. Then from the displayed equation, p divides $10b$. Since we are assuming that a and b are relatively prime we cannot have that p divides b, but it could divide 10. (In fact this happens in the example of the last theorem where $a = 22$; each remainder there is divisible by 2 also.)

Evidently if we want r_1 and a to be relatively prime we will have to assume that no prime dividing a also divides 10; that is, we must assume that $(a, 10) = 1$. Then the above argument leads to a contradiction, forcing us to the conclusion that $(a, r_1) = 1$.

That all the remainders r_i are relatively prime to a now follows in the same way. For example we have

$$10r_1 = q_2a + r_2,$$

and so any prime dividing a and r_2 would have to divide either 10 or r_1, which we know cannot happen. Clearly the argument may be continued indefinitely.

It may be a good idea to recap what we have figured out so far.

LEMMA   Let b be divided by a by the long-division algorithm, where $0 \leq b < a$, and $(a, b) = 1$. Then the remainders r_i satisfy

$$r_{i+j} \equiv 10^j r_i \pmod{a}$$

for all positive integers i and j. If $(a, 10) = 1$, then all the remainders are relatively prime to a.

Recall that in the first theorem of this section we saw that the digits in the decimal expansion of b/a repeat whenever a remainder repeats itself. Suppose that j is the period of the expansion of b/a, so that for some i we have $r_i = r_{i+j}$. Then we have

$$r_i \equiv 10^j r_i \pmod{a}.$$

If $(a, 10) = 1$ we know that r_i is relatively prime to a and so can be cancelled by the Cancellation Theorem. This gives

$$10^j \equiv 1 \pmod{a}.$$

Furthermore, since we are assuming that j is the period of the expansion of b/a, no smaller positive power of 10 can be congruent to 1 (mod a).

In investigating the period of a rational number we have come back to a concept first defined in Section 4.4, namely the order of one integer modulo another. Recall that if $(x, a) = 1$, then the order of x (mod a) is the least positive integer k such that

$$x^k \equiv 1 \pmod{a}.$$

In the present case we have found that j, which is the period of the decimal expansion of b/a, is also the order of 10 (mod a).

A pertinent result is Theorem 4.5, which says that the order of an integer modulo a must divide $\varphi(a)$. Applying this theorem with $x = 10$ gives the last line of the following result.

THEOREM   Let the decimal expansion of the rational number b/a have period j, where $0 \le b < a$, $(a, b) = 1$, and $(a, 10) = 1$. Then j is the order of 10 modulo a. In particular, j divides $\varphi(a)$.

Example   Find the period of the decimal expansion of 2/7.

Although we could just divide 7 into 2, we will instead compute the order of 10 (mod 7). This must be a divisor of $\varphi(7) = 6$. We have

$$10 \equiv 3 \quad (\text{mod } 7), \qquad 10^2 \equiv 9 \equiv 2 \quad (\text{mod } 7), \qquad \text{and}$$
$$10^3 \equiv 3 \cdot 2 \equiv 6 \quad (\text{mod } 7).$$

Since the order of 10 (mod 7) is not 1, 2, or 3, it must be 6. In fact $2/7 = .\overline{285714}$.

Notice that the period of the expansion of b/a depends only on the denominator a (assuming $(a, b) = 1$, of course).

Decimals for 1/n with Maximal Period

People who play around with numbers have often considered the question of when the period of the decimal representation of $1/n$ has length $n - 1$, as, for example, $1/7 = .\overline{142857}$, which has period 6. Since we know (assuming that $(n, 10) = 1$) the period must be a divisor of $\varphi(n)$, which is $n - 1$ only when n is prime, this happens only for prime values of n. Having n prime is not sufficient for $1/n$ to have period $n - 1$, however. For example $1/11 = .090909 \ldots = .\overline{09}$, which has period 2. This is to be expected, since $10 \equiv -1$ (mod 11), so the order of 10 (mod 11) is 2.

In general, if $(g, n) = 1$, then the order of g (mod n) is a divisor of $\varphi(n)$; the question is when it actually equals $\varphi(n)$.

DEFINITION   primitive root

Let n be a positive integer and suppose $(g, n) = 1$. We say that g is a **primitive root** modulo n in case the order of g (mod n) is $\varphi(n)$.

Examples   We have seen that 10 is a primitive root (mod 7) but not a primitive root (mod 11). The number 11 does have a primitive root, however, namely $g = 2$. Since the order of 2 (mod 11) must be a divisor of $\varphi(11) = 10$, in order to check this it suffices that we compute that neither 2, $2^2 = 4$, nor $2^5 = 32$ is congruent to 1 (mod 11).

Some integers have no primitive roots. For example if $n = 8$ the only candidates are $g = 1, 3, 5$, and 7. Of course the order of 1 is 1, while the other three numbers have order 2. No integer has order $\varphi(8) = 4$.

From the above discussion if $(n, 10) = 1$ the decimal for $1/n$ has period $n - 1$ if and only if n is prime and 10 is a primitive root (mod n). Unfortunately the latter characterization is not really helpful, and not much is known about which primes have 10 as a primitive root. It is not even known if there are infinitely many such primes.

PROBLEMS FOR SECTION 6.4

In the first eight problems write the given rational number as a periodic decimal.

1. 8/33   2. 13/27   3. 5/14   4. 17/81
5. 6/13   6. 221/37   7. 81/52   8. 2/105

In the next eight problems write the given periodic decimal in the form b/a with $(a, b) = 1$.

9. $.\overline{7}$ 10. $.0\overline{5}$ 11. $.7\overline{3}$ 12. $.\overline{171}$
13. $1.3\overline{5}$ 14. $2.29\overline{54}$ 15. $.\overline{4356}$ 16. $2.0\overline{79002}$

In the next six problems determine the period of the decimal representation of the given number.

17. 93/7   18. 100/13   19. 112/21
20. 17/33   21. 3/101   22. 13/77

In the next six problems for each given prime p determine whether the decimal for $1/p$ has period $p - 1$ or not.

23. 13   24. 17   25. 19   26. 23   27. 29   28. 31

In the next six problems determine the smallest positive integer g, if any, such that g is a primitive root modulo the given integer.

29. 10   30. 12   31. 14   32. 15   33. 16   34. 18

35. Find all positive integers a, $(a, 10) = 1$, such that the decimal expansion of $1/a$ has period 2.
36. Find all positive integers a, $(a, 10) = 1$, such that the decimal expansion of $1/a$ has period 3.
37. Show that if b, r, and s are nonnegative integers, then $b/2^r 5^s$ has a terminating decimal expansion.
38. Show that if b/a has a terminating expansion, where a and b are positive integers and $(a, b) = 1$, then $a = 2^r 5^s$ for nonnegative integers r and s.

39. Show that if $(a, b) = (a, b') = 1$, with a, b, and b' positive integers, and if

$$b/a = e_1 \cdots e_m.d_1 \cdots d_k\overline{d_{k+1} \cdots d_{k+j}},$$

then there exist digits $f_1, \ldots, f_r, g_1, \ldots, g_s$ such that

$$b'/a = f_1 \cdots f_r.g_1 \cdots g_s\overline{d_{k+1} \cdots d_{k+j}}.$$

40. Let a and b be positive integers with $(a, b) = 1$. Suppose $a = 2^r5^sa'$, where $(a', 10) = 1$. Show that the period of a decimal expansion of b/a is the order of $10 \pmod{a'}$.

41. Suppose $(10, m) = 1$. Show that there exists an integer k such that the decimal representation of k consists only of 9's, and such that $m|k$. (**Hint:** Consider $10^j \equiv 1 \pmod m$.)

42. Suppose $(10, m) = 1$. Show that there exists an integer k such that the decimal representation of k consists only of 1's, and such that $m|k$.

C *The next three problems concern b-ary expansions, that is expansions of the form*

$$.a_1a_2 \cdots = \frac{a_1}{b} + \frac{a_2}{b^2} + \cdots,$$

where $b > 1$ and $0 \le a_i < b$ for all i.

43. Show that a b-ary expansion of $1/n$ terminates if and only if n divides some integral power of b.

44. Show that if $(n, b) = 1$ then the period of a b-ary expansion of $1/n$ equals the order of $b \pmod n$.

45. Show that if a positive real number has more than one b-ary expansion, then it has exactly 2, and one of these is terminating.

6.5 Lagrange's Theorem and Primitive Roots for Prime Modulus

The main purpose of this section is to prove that if p is prime, then there exists a primitive root modulo p, that is, an element of order $\varphi(p) = p - 1 \pmod p$. We know that if $(g, p) = 1$, then g has some order $k \pmod p$, and that k divides $p - 1$. In the course of our proof we will actually determine how many elements any reduced residue system $\pmod p$ has of each possible order.

Let us denote by $N(k)$ the number of elements in a reduced residue system $\pmod p$ of order k. For example in Section 4.4 we saw that for $p = 7$ the numbers 1, 2, 3, 4, 5, and 6 have orders 1, 3, 6, 3, 6, and 2, respectively. Thus in this case $N(1) = 1$, $N(2) = 1$, $N(3) = 2$, and $N(6) = 2$. Of course if k does not divide $p - 1$ then $N(k) = 0$. Since a reduced residue system $\pmod p$ has $p - 1$ elements, each of which has some order dividing $p - 1$, we have

$$\sum_{k|p-1} N(k) = p - 1. \tag{1}$$

In our example $N(1) + N(2) + N(3) + N(6) = 1 + 1 + 2 + 2 = 6 = 7 - 1$.

It turns out that $N(k)$ cannot be too large. For example always $N(1) = 1$, since only the element of a reduced residue system congruent to 1 (mod p) can have order 1.

How many elements can have order 2? If g has order 2 (mod p), then $g^2 \equiv 1$ (mod p). This means $p | g^2 - 1 = (g - 1)(g + 1)$. Since p must divide one or the other of these factors, either $g \equiv 1$ or $g \equiv -1$ (mod p). Since 1 has order 1 and -1 has order 2 (mod p) for $p > 2$, we have $N(2) = 1$. (If $p = 2$ then $-1 \equiv 1$ and we have $N(2) = 0$. Note that the argument depends upon the fact that p is prime. The integers 3, 5, and 7 all have order 2 (mod 8).)

In the same way if g has order 3 (mod p), then $g^3 \equiv 1$ (mod p), so p divides $g^3 - 1 = (g - 1)(g^2 + g + 1)$. Thus if g is not congruent to 1 (mod p) it must satisfy the quadratic congruence

$$x^2 + x + 1 \equiv 0 \quad (\text{mod } p).$$

By the first theorem of Section 5.3 this congruence has at most 2 solutions. We see that $N(3)$ is at most 2.

Lagrange's Theorem

In general if g has order k (mod p), then g satisfies

$$x^k - 1 \equiv 0 \quad (\text{mod } p).$$

We would like to be able to say that this congruence, involving a polynomial of degree k, has at most k solutions in any complete residue system (mod p), just as from algebra we know that a polynomial equation of degree k has at most k solutions. First it is necessary to build up a small amount of theory about polynomials.

LEMMA 1   Let f be a polynomial with integer coefficients, and let r be any integer. Then there exists a polynomial g with integer coefficients such that

$$f(x) - f(r) = (x - r)g(x).$$

Proof   Let $f(x) = a_n x^n + a_{n-1} x^{n-1} + \cdots + a_0$. Then

$$f(x) - f(r) = a_n(x^n - r^n) + a_{n-1}(x^{n-1} - r^{n-1}) + \cdots + a_1(x - r)$$

$$= (x - r)g(x),$$

since it is easily seen that for each positive integer j

$$x^j - r^j = (x - r)(x^{j-1} + x^{j-2}r + x^{j-3}r^2 + \cdots + r^{j-1})$$

by multiplying out the right side and cancelling.

LEMMA 2   If f is a nonconstant polynomial with integer coefficients and $f(r) \equiv 0$ (mod m), then there exists a polynomial g with integer coefficients with the same leading coefficient as f such that the corresponding coefficients of $f(x)$ and $(x - r)g(x)$ are all congruent (mod m) and such that the degree of g is one less than that of f.

Proof   From Lemma 1 there exists an integral polynomial g such that

$$f(x) - f(r) = (x - r)g(x).$$

Clearly g has the same leading coefficient as f and degree one less. The coefficients of $(x - r)g(x)$ are the same as those of $f(x)$ except for the constant terms, which differ by $f(r) \equiv 0$ (mod m).

If two integral polynomials P_1 and P_2 have corresponding coefficients congruent (mod m), we will write

$$P_1(x) \sim P_2(x) \quad \text{(mod } m\text{)}.$$

For example in Lemma 2 we have $f(x) \sim (x - r)g(x)$ (mod m).

LAGRANGE'S THEOREM   Let f be a polynomial with integer coefficients of degree n, and let p be a prime. Suppose p does not divide the leading coefficient of f. Then the congruence

$$f(x) \equiv 0 \quad \text{(mod } p\text{)} \tag{$*$}$$

has at most n solutions in any complete residue system (mod p).

Proof   Suppose $r_1, r_2, \ldots, r_{n+1}$ are incongruent solutions of ($*$). By Lemma 2 we can write

$$f(x) \sim (x - r_1)g(x) \quad \text{(mod } p\text{)},$$

where g has the same leading coefficient as f but degree 1 less. Then for $i = 2, 3, \ldots, n + 1$

$$f(r_i) \equiv (r_i - r_1)g(r_i) \equiv 0 \quad \text{(mod } p\text{)}.$$

Since r_i is not congruent to r_1 (mod p) for $i > 1$, we see that $r_2, r_3, \ldots, r_{n+1}$ all satisfy

$$g(x) \equiv 0 \quad (\text{mod } p).$$

Thus in the same way

$$g(x) \sim (x - r_2)h(x) \quad (\text{mod } p),$$

where h has the same leading coefficient as g and degree one less. This means

$$f(x) \sim (x - r_1)(x - r_2)h(x) \quad (\text{mod } p).$$

If we continue in this way we get

$$f(x) \sim (x - r_1) \cdots (x - r_n)c \quad (\text{mod } p), \qquad (**)$$

where c is the leading coefficient of f. But then $f(r_{n+1}) \equiv 0 \,(\text{mod } p)$ is impossible, since when r_{n+1} is substituted into $(**)$ for x none of the factors on the right is divisible by p.

The Number of Elements of a Given Order Modulo p

In the light of Lagrange's Theorem we see that since any element of order k (mod p) satisfies

$$x^k \equiv 1 \quad (\text{mod } p), \qquad (2)$$

there can be at most k such elements in any reduced residue system (mod p), and so $N(k) \leq k$. Actually we can do a little better than this. Suppose $N(k)$ is not 0, so there exists an integer a with order k (mod p). From Theorem 4.4 we know that the elements $a, a^2, \ldots, a^k$ are all distinct (mod p). It is easy to see that each of them also satisfies (2), since

$$(a^j)^k = (a^k)^j \equiv 1^j \equiv 1 \quad (\text{mod } p).$$

Since Lagrange's Theorem says that (2) can have at most k distinct solutions (mod p), the numbers $a, a^2, \ldots, a^k$ must constitute all of them.

Not all of these powers of a have order k, however. We claim that a^j has order k (mod p) if and only if $(j, k) = 1$. To see this first suppose that $(j, k) = d > 1$. Then

$$(a^j)^{k/d} = (a^k)^{j/d} \equiv 1^{j/d} \equiv 1 \quad (\text{mod } p),$$

from which we see that the order of a^j does not exceed $k/d < k$.

If $(j, k) = 1$, on the other hand, we will show that for no positive integer $h < k$ do we have

$$(a^j)^h \equiv 1 \pmod{p}. \tag{3}$$

For if we suppose that (3) holds, then by Theorem 4.4 we have $k \mid jh$. But since $(j, k) = 1$ this means that $k \mid h$. Thus $h \geq k$.

Since we have proved that if a has order k then the elements of order k in some reduced residue system (mod p) are exactly the integers a^j, $1 \leq j \leq k$ with $(j, k) = 1$, we have established the following result.

LEMMA   Suppose p is prime and k divides $p - 1$. Then $N(k)$ is either 0 or $\varphi(k)$.

Example   The reader should check that the following table correctly gives the order of each of the integers $1, 2, \ldots, 12 \pmod{13}$.

element	1	2	3	4	5	6	7	8	9	10	11	12
order	1	12	3	6	4	12	12	4	3	6	12	2

The computation of orders can be simplified by what we learned in the proof of the lemma. For example, once we know that the order of 2 (mod 13) is 12, then the elements $6 \equiv 2^5$, $11 \equiv 2^7$, and $7 \equiv 2^{11}$ must also have order 12 (and so be primitive roots modulo 13), since 5, 7, and 11 are relatively prime to 12.

In this case we have $N(1) = 1$, $N(2) = 1$, $N(3) = 2$, $N(4) = 2$, and $N(6) = 2$. Notice that for each divisor k of $\varphi(13) = 12$ we have $N(k) = \varphi(k)$. We will prove that this is always the case.

From the last Lemma and the formula (1) we can now say that

$$p - 1 = \sum N(k) \leq \sum \varphi(k), \tag{4}$$

where both summations run over the positive divisors k of $p - 1$. By Theorem 3.5 the summation on the right is exactly $p - 1$, and so we have

$$p - 1 = \sum_{k \mid p-1} N(k) \leq \sum_{k \mid p-1} \varphi(k) = p - 1.$$

Since we have $p - 1$ on both extremes, all the inequalities must be equalities; that is, $N(k) = \varphi(k)$ whenever k divides $p - 1$.

THEOREM 6.5   Suppose p is prime and k divides $p - 1$, $k > 0$. Then any reduced residue system modulo p contains exactly $\varphi(k)$ elements having order $k \pmod{p}$. In particular, there exist primitive roots modulo p.

Notice that this is merely an existence theorem, and does not give any way to find a primitive root for a particular prime. It has not even been proved that there is any fixed number, such as 2, which is a primitive root for infinitely-many primes, although this seems likely.

Moduli Having Primitive Roots

It can be shown that the modulus n has a primitive root exactly in the following cases:

a. $n = p^r$ for some odd prime p,
b. $n = 2p^r$ for some odd prime p,
c. $n = 2$ or $n = 4$.

For example, among the integers from 20 to 30, precisely 23, 25, 27, 29 (all case (a)), 22, and 26 (both case (b)) have primitive roots. This statement will not be proved here, although some cases will be found in the exercises for this section.

Indices

Suppose g is a primitive root modulo n. Then the powers of g run through a reduced residue system (mod n). This means that elements of this reduced residue system can be multiplied by adding the corresponding exponents of g.

As an example we take $n = 13$ and $g = 2$. The following table gives the least residue of each number 2^j, $j = 0, 1, \ldots, 11$.

j	0	1	2	3	4	5	6	7	8	9	10	11
2^j (mod 13)	1	2	4	8	3	6	12	11	9	5	10	7

DEFINITION   index
If g is a primitive root (mod n) and $(a, n) = 1$, by the **index** of a we mean the number j, $0 \le j < \varphi(n)$, such that $g^j \equiv a$ (mod n). (Of course the index of a depends on n and g also.) For example, from the above table we see that the index of 8 is 3, and the index of 3 is 4.

It is useful to turn our table around so as to display the index of each number.

a	1	2	3	4	5	6	7	8	9	10	11	12
index of a	0	1	4	2	9	5	11	3	8	10	7	6

This table of indices can be used to multiply (mod 13) as follows. Suppose we want the least residue of $8 \cdot 11$ (mod 13). We have that

$$8 \cdot 11 \equiv 2^3 2^7 \equiv 2^{10} \equiv 10 \pmod{13}.$$

In the same way we compute that

$$5 \cdot 9 \cdot 7 \equiv 2^9 2^8 2^{11} \equiv 2^{28} \equiv (2^{12})^2 2^4 = 1 \cdot 3 \equiv 3 \pmod{13},$$

where at the end we used the fact that $2^{12} \equiv 1$ (mod 13) by Euler's Theorem. This means that exponents may be reduced modulo 12 (or, in general, modulo $\varphi(n)$).

This system replaces multiplication by addition and works for reduced residues (mod n) exactly as logarithms do for real numbers. Likewise division (mod n) may be replaced by subtraction; but we leave the details for the reader to work out.

PROBLEMS FOR SECTION 6.5

 In the first eight problems for each given number n tell how many primitive roots there are in a reduced residue system (mod n).

1. 17	2. 19	3. 10	4. 14
5. 101	6. 103	7. 105	8. 107

9. Which of the numbers 50, 51, . . . , 59 have primitive roots?
10. Which of the numbers 80, 81, . . . , 89 have primitive roots?

In the next four problems use the tables for $p = 13$ at the end of this section to find the least residue (mod 13) of the given number.

11. $9 \cdot 12$	12. $11 \cdot 7$	13. $6 \cdot 7 \cdot 8$	14. $5 \cdot 7 \cdot 9$

15. Make a table like the last one in this section for the modulus 7 and primitive root 3.
16. Make a table like the last one in this section for the modulus 11 and primitive root 6.

 True-False. In the next nine problems tell which statements are true, and give counterexamples for those that are false. Assume that a and b are primitive roots (mod m).

17. ab is a primitive root (mod m).
18. $-a$ is a primitive root (mod m).
19. $a + b$ is a primitive root (mod m).
20. $a + b$ is not a primitive root (mod m).
21. a^2 is not a primitive root (mod m).
22. If $b \equiv a^k$ (mod m), then $(m, k) = 1$.
23. If $m > 2$, then $a^{\varphi(m)/2} \equiv -1$ (mod m).
24. If c is a solution to $ac \equiv 1$ (mod m), then c is a primitive root (mod m).
25. If $c^{\varphi(m)/2} \equiv -1$ (mod m), then c is a primitive root (mod m).

26. Let p be an odd prime and g a primitive root (mod p). Prove g is not a quadratic residue (mod p).
27. Show that if g is a primitive root (mod p), p prime, and $(a, p) = 1$, then there exists a unique t, $0 \le t < p - 1$, such that $g^t \equiv a$ (mod p).

28. Let g, a, p, and t be as in the last problem. Show that a is a quadratic residue (mod p) if and only if t is even.

29. Show that Lagrange's Theorem does not hold if the modulus is not restricted to be prime.

30. Show that Lagrange's Theorem does not hold if the restriction that the leading coefficient of f is not divisible by p is dropped.

31. Show that if F_r is prime, then any quadratic nonresidue (mod F_r) is a primitive root. (See Section 3.4.)

32. Let a and b be positive integers with $(a, b) = 1$. Show that if g is a primitive root (mod ab), then g is a primitive root modulo both a and b.

33. Show that the only powers of 2 having primitive roots are 2 and 4.

C 34. Let p and $q = 2p + 1$ be primes, with $p \equiv 3$ (mod 4). Show that every quadratic nonresidue (mod q) not congruent to -1 (mod q) is a primitive root (mod q).

35. With the hypotheses of the last problem, show that -2 is a primitive root for q.

36. Suppose F_r is prime, $r > 0$. (See Section 3.4.) Show that 3 is a primitive root (mod F_r). (**Hint:** Show $F_r \equiv 5$ (mod 12).)

37. Let g be a primitive root (mod n), n odd. Let h be g or $g + n$, whichever is odd. Show that h is a primitive root (mod $2n$).

38. Let $p \equiv 1$ (mod 4) be prime. Show that if g is a primitive root (mod p), then so is $-g$.

39. Let $(a, b) = 1$ and suppose g has order r (mod a) and s (mod b). Show that the order of g (mod ab) is $[r, s]$.

40. Show that if ab has a primitive root, with $(a, b) = 1$, then $a < 3$ or $b < 3$. (**Hint:** Use the last problem to show that $\varphi(a)$ and $\varphi(b)$ are relatively prime.)

41. Show that if n has a primitive root and if the odd prime p divides n, then n is of the form p^j or $2p^j$.

G.H. Hardy and Srinivasa Ramanujan
1877–1947   1887–1920

Hardy hated to have his picture taken and used towels to cover all the mirrors in hotel rooms to avoid seeing his own image. He called God his personal enemy, refusing to enter any religious building. Ramanujan was a vegetarian who always changed into pajamas before cooking his own meals. He agreed to travel from India to England only after his mother had a dream in which the goddess Namagiri approved the journey. The personal idiosyncrasies of Hardy and Ramanujan are no more remarkable, however, than the mathematical chain of events that brought them together.

G.H. HARDY

G.H. Hardy was a Cambridge professor who defended his colleague Bertrand Russell's antiwar activities during World War I. Hardy's specialty was analytic number theory, an area in which he and his collaborator, J.E. Littlewood, made fundamental advances. Hardy was a witty conversationalist who loved tennis and cricket, and adopted baseball when visiting the United States. His small book *A Mathematician's Apology* lays out in graceful prose his life in the purest of pure mathematics.

One day in 1913 Hardy received a letter from India full of mathematical formulas. Some were standard results, but others were like nothing he had seen before. Hardy recognized the genius of the writer of the letter, Srinivasa Ramanujan, and immediately arranged for him to come to Cambridge.

SRINIVASA RAMANUJAN

Ramanujan was born in the Tanjore district of Madras, the son of an accountant to a cloth merchant. At 15, Ramanujan gained access to Carr's *Synopsis of Pure Mathematics*, a compendium of formulas of elementary mathematics. Ramanujan used this book to teach himself the subject, and soon he was discovering mathematical laws on his own. Lacking formal mathematical training, he proceeded on a basis of intuition and experiment that is hard to reconstruct. According to Hardy, Ramanujan lacked the modern idea of a proof.

When Ramanujan got to England, Hardy found in him a genius who was ignorant of many fundamental areas of mathematics and had the problem of acquainting him with modern ideas without stifling his originality. They wrote a series of papers together.

The British climate did not agree with Ramanujan, and he developed tuberculosis. When Hardy visited him in a hospital to make conversation he mentioned that the number of his taxicab had been 1729, "a rather dull number." Ramanujan replied "No, Hardy! No, Hardy! It is a very interesting number. It is the smallest number expressible as the sum of two cubes in two different ways." Ramanujan returned to India for his health, but died soon after, at the age of 32.

Chapter 7

Diophantine Equations

Although **Diophantine equations** are named for the Greek mathematician Diophantus, who generally sought rational solutions, we will use the term in its more common modern sense, referring to equations for which integral solutions are desired. Almost nothing is known about the personal life of Diophantus, other than that he lived in Alexandria, perhaps about 250 A.D.; but those of his books that survived set a mathematical standard unsurpassed for many centuries. (See the problem at the end of Section 2.3 for more information about him.) The achievements of Diophantus are even more remarkable when one considers that he did not have the advantage of modern algebraic notation, and so had to describe his methods in words or by means of numerical examples.

We have already considered some Diophantine equations, namely the equation $ax + by = c$ of Section 1.4, and the equation $x^n + y^n = z^n$ of Fermat's Last Theorem (Section 2.3). Actually Diophantus never seems to have written about linear equations such as the former, perhaps because he thought them too simple.

There are few general theorems concerning Diophantine equations, and special proof techniques are often necessary for each equation. We will consider only a few of the simpler and more interesting examples of Diophantine equations.

7.1 Pythagorean Triples

DEFINITIONS Pythagorean triple, primitive Pythagorean triple

We call the triple of positive integers (x, y, z) a **Pythagorean triple** in case it satisfies the Diophantine equation

$$x^2 + y^2 = z^2. \tag{$*$}$$

The triple is said to be **primitive** in case no integer greater than 1 divides all of x, y, and z.

Of course by the Pythagorean Theorem in any right triangle whose sides have integral lengths, these lengths form a Pythagorean triple; and by its converse any such triple corresponds to a right triangle. This accounts for the terminology. The best known Pythagorean triple is (3, 4, 5), which is primitive. Other (nonprimitive) triples such as (6, 8, 10), (9, 12, 15), and (30, 40, 50) can easily be formed from this one, since it is easy to see that if (x, y, z) is a Pythagorean triple, then so is (kx, ky, kz) for any positive k.

We see that any primitive Pythagorean triple (x, y, z) generates infinitely many nonprimitive ones (kx, ky, kz), $k = 2, 3, \ldots$. Conversely, any Pythagorean triple (x, y, z) corresponds to a unique primitive one found by dividing x, y, and z by their greatest common divisor. Thus we will concentrate our attention on primitive triples.

Other primitive Pythagorean triples exist, for example (12, 5, 13), which the reader should check. In fact there are infinitely many. This can be demonstrated by the following simple geometric argument. Let n be any odd positive integer. Then n^2 is also odd, and so of the form $2m + 1$. Arrange $n^2 = 2m + 1$ unit squares outside two sides of an m-by-m square as follows (in the example we have taken $n = 3$ so $2m + 1 = 9$ and $m = 4$):

```
*   *   *   *   *
*   O   O   O   O
*   O   O   O   O
*   O   O   O   O
*   O   O   O   O
```

Then (referring to areas)

$$(\text{inner square}) + (\text{border}) = (\text{large square}), \qquad \text{or}$$

$$m^2 \quad + \quad n^2 \quad = \quad (m + 1)^2.$$

This shows that $(m, n, m + 1)$ is a Pythagorean triple, primitive since m and $m + 1$ are relatively prime. Since n was an arbitrary odd positive integer, there exist infinitely many such triples.

Examples The case illustrated above leads to the triple (4, 3, 5), while taking $n = 5$ makes $2m + 1 = 25$ and $m = 12$. This gives the already-mentioned triple (12, 5, 13).

Finding All Primitive Solutions

One might guess that all primitive Pythagorean triples arise from the above method, but such is not the case. For example (8, 15, 17) is Pythagorean, since

$64 + 225 = 289 = 17^2$. Yet taking $n = 15$ in the geometric method makes $2m + 1 = 225$ and $m = 112$, leading to the triple $(112, 15, 113)$ instead. We will find a slightly more complicated way of generating all Pythagorean triples in this section.

LEMMA   If (x, y, z) is a primitive Pythagorean triple, then x, y, and z are relatively prime in pairs, one of x and y is even and the other odd, and z is odd.

Proof   Suppose $(x, y) = d$. If $d > 1$ then some prime p divides d. From equation (∗) and Theorem 1.2 we see that p divides z^2. But then by the Fundamental Theorem of Arithmetic p divides z also, contradicting the assumption that (x, y, z) is primitive. Similar proofs show that $(x, z) = (y, z) = 1$ also.

Now from the first part of this lemma x and y cannot both be even. We claim that they cannot both be odd either. For if they were, then (∗) would imply that z was even. But if x and y are odd and z even, we have

$$x \equiv 1 \text{ or } 3 \pmod 4,$$

$$y \equiv 1 \text{ or } 3 \pmod 4, \qquad \text{and}$$

$$z \equiv 0 \text{ or } 2 \pmod 4.$$

Squaring these congruences gives

$$x^2 \equiv 1 \pmod 4,$$

$$y^2 \equiv 1 \pmod 4,$$

$$z^2 \equiv 0 \pmod 4.$$

But adding the first two congruences gives

$$x^2 + y^2 \equiv 2 \pmod 4,$$

which contradicts (∗) and the third congruence. Thus one of x and y is even and the other odd, and z must also be odd.

Now we will turn to a general analysis of primitive Pythagorean triples. We have seen from the lemma that one of x and y is even and the other odd. Since (∗) is symmetric in x and y we may label a solution so that x is even, say $x = 2t$, and y is odd. Substituting this equation in (∗) gives

$$4t^2 + y^2 = z^2, \qquad \text{or}$$

$$4t^2 = z^2 - y^2 = (z + y)(z - y).$$

Since y and z are both odd the two factors on the right are even, so we can write

$$t^2 = \frac{z+y}{2} \cdot \frac{z-y}{2}, \tag{$**$}$$

where the factors on the right are not only integers, but relatively prime, since any common factor greater than 1 would have to divide both their sum, which is z, and their difference, which is y, contradicting the lemma.

Now since each prime dividing the left side of the last equation does so to an even power, the same goes for the right side of the equation. Because no prime can divide both of the right-hand factors, each contains only even prime powers and so is a perfect square, say

$$\frac{z+y}{2} = u^2 \quad \text{and} \quad \frac{z-y}{2} = v^2,$$

where we may assume u and v are positive.

Adding these two equations gives $z = u^2 + v^2$, and subtracting them gives $y = u^2 - v^2$, from which we see that $u > v$ since $y > 0$. Note that since z is odd u and v cannot be both even or both odd. Since we saw above that u^2 and v^2 were relatively prime, the same goes for u and v.

Finally from $(**)$

$$t^2 = u^2 v^2,$$

and so $x = 2t = 2uv$.

THE PYTHAGOREAN TRIPLE THEOREM   Suppose that (x, y, z) is a primitive Pythagorean triple with x even. Then there exist relatively prime positive integers u and v, not both odd, with $u > v$, such that

$$x = 2uv, \quad y = u^2 - v^2, \quad \text{and} \quad z = u^2 + v^2.$$

Conversely, if u and v are any pair of integers as described in the previous sentence, and if x, y, and z are defined by these three equations, then (x, y, z) is a primitive Pythagorean triple with x even.

Proof   The first part of the theorem has just been proved. To prove the second part, assume that u and v are as described, and let x, y, and z be defined by the three equations. Then

$$x^2 + y^2 = (2uv)^2 + (u^2 - v^2)^2$$

$$= 4u^2 v^2 + u^4 - 2u^2 v^2 + v^4$$

$$= u^4 + 2u^2 v^2 + v^4 = (u^2 + v^2)^2 = z^2,$$

and so $(*)$ is satisfied.

Clearly x is even. It only remains to show that (x, y, z) is primitive. Suppose some prime p divides both y and z. Then it divides their sum, which is $2u^2$, and their difference, which is $2v^2$. Since by hypothesis one of u and v is even and the other odd, y is odd and so p is not 2. Then p divides both u and v, contradicting the assumption that u and v are relatively prime.

Examples The following table gives Pythagorean triples corresponding to some small values of u and v.

u	v	x	y	z
2	1	4	3	5
3	2	12	5	13
4	1	8	15	17
4	3	24	7	25

Note that any pair of positive integers u and v will generate a Pythagorean triple (x, y, z); the extra conditions on u and v are necessary to guarantee that the triple is primitive.

Example A Find all primitive Pythagorean triples with $x = 28$.
    Since $x = 2uv$, we have $uv = 14$. The only possibilities for u and v satisfying the theorem are $u = 14$, $v = 1$, and $u = 7$, $v = 2$. These lead to the triples $(28, 195, 197)$ and $(28, 45, 53)$, respectively.

Example B Find all primitive Pythagorean triples with $y = 15$.
    We need to solve the equation $u^2 - v^2 = 15$. Notice that there are only finitely many solutions possible, since the difference between consecutive squares gets larger and larger.
    In fact $15 = (u + v)(u - v)$, and the factors on the right must be either 15 and 1 or else 5 and 3. Solving the equations

$$u + v = 15, \qquad u - v = 1$$

gives $u = 8$ and $v = 7$, leading to the triple $(112, 15, 113)$; while the other case gives $u = 4$ and $v = 1$, leading to $(8, 15, 17)$.

Example C Find all Pythagorean triples with $x = 28$.
    This looks just like Example A until we notice that the condition that the triples be primitive has been dropped. A nonprimitive solution will be a triple $(28, y, z) = (ka, kb, kc)$, where (a, b, c) is primitive and a is some divisor of 28. We cannot even be sure that a is even.
    The divisors of 28 are 1, 2, 4, 7, 14, and 28, and we will try each of these in turn as elements of a primitive triple. Let us start with the even divisors, which

would correspond to *2uv* in a primitive triple. The equation $2uv = 2$ leads to $uv = 1$, which is impossible since *u* and *v* are not both odd. The equation $2uv = 14$ is impossible in a similar way. However $2uv = 4$ says $uv = 2$, which works with $u = 2$ and $v = 1$. This yields the primitive triple (4, 3, 5). We multiply this by 7 to get the nonprimitive triple (28, 21, 35). The equation $2uv = 28$ was already solved in Example A, yielding the primitive triples (28, 195, 197) and (28, 45, 53).

Now let us consider the odd divisors of 28. To appear as *y* in a primitive solution they would have to be of the form $u^2 - v^2$. The equation $1 = u^2 - v^2$ is impossible, but $7 = u^2 - v^2 = (u + v)(u - v)$ has the solution $u = 4$, $v = 3$, leading to the primitive triple (24, 7, 25). We multiply this by 4 to get the nonprimitive triple (96, 28, 100). Since it was *x* that was supposed to be 28 we rearrange this to (28, 96, 100).

We have found four answers to our problem, namely (28, 21, 35), (28, 195, 197), (28, 45, 53), and (28, 96, 100).

PROBLEMS FOR SECTION 7.1

 The first four problems refer to the geometric method given at the beginning of this section. What Pythagorean triples are generated by the following values of n?

   1. $n = 7$ 2. $n = 9$ 3. $n = 11$ 4. $n = 13$

In the next four problems find the triples (x, y, z) generated by the given u and v.

   5. $u = 6, v = 1$ 6. $u = 6, v = 5$
   7. $u = 5, v = 2$ 8. $u = 5, v = 4$

In the next four problems find the u and v generating the given triple (x, y, z).

   9. (24, 7, 25)   10. (48, 55, 73)
  11. (28, 45, 53)   12. (140, 51, 149)

In the next seven problems assume that (x, y, z) is a primitive Pythagorean triple generated by u and v. Find the three missing variables.

  13. $x = 84, y = 13$ 14. $u = 7, x = 56$ 15. $u = 9, y = 17$
  16. $v = 4, z = 97$ 17. $u = 8, z = 89$ 18. $v = 7, y = 15$
  19. $v = 2, x = 36$

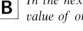 *In the next seven problems find all primitive Pythagorean triples with the given value of one variable.*

  20. $x = 12$ 21. $x = 24$ 22. $y = 9$ 23. $y = 21$
  24. $y = 20$ 25. $z = 125$ 26. $z = 65$

In the next four problems find all Pythagorean triples with the given value of one variable.

27. $x = 16$ 28. $x = 40$ 29. $y = 35$ 30. $z = 85$

31. Show that if (x, y, z) is a primitive Pythagorean triple, then $(x, z) = (y, z) = 1$.
32. Show that just because no integer greater than 1 divides all 3 of x, y, and z, it need not be true that x, y, and z are relatively prime in pairs.
33. Each solution by the geometric method given at the beginning of this section corresponds to a solution generated by u and v as in the Pythagorean Triple Theorem. What are u and v in terms of the odd number n?

$\boxed{\textbf{C}}$ 34. Show that if $x \equiv 2 \pmod 4$, then there is no primitive Pythagorean triple (x, y, z).
35. Suppose a circle is inscribed in a right triangle whose sides have integral lengths. Show that the radius of this circle is an integer.

In the next five problems assume that x, y, and z are positive integers satisfying

$$x^2 + 2y^2 = z^2, \qquad\qquad\qquad (\ast\ast\ast)$$

and that no integer greater than 1 divides all of x, y, and z. These problems parallel the proof of the Pythagorean Triple Theorem (with some variations).

36. Show that x, y, and z are relatively prime in pairs.
37. Show that x and z are odd and y is even.
38. Show that the integers $(z + x)/2$ and $(z - x)/2$ are relatively prime, and exactly one of them is even.
39. Define u to be the square root of half of whichever of the two integers in the last problem is even, and v to be the square root of whichever of them is odd. Then u and v are integers and $(2u, v) = 1$.
40. We have $x = |2u^2 - v^2|$, $y = 2uv$, $z = 2u^2 + v^2$.

41. If x, y, and z are defined by the equations of the previous problem, then they satisfy $(\ast\ast\ast)$. Furthermore, if $(2u, v) = 1$, then x, y, and z have no common divisor greater than 1.

7.2 Sums of Two Squares

In the last section we saw that for each pair of positive integers u, v with $u > v$, defining

$$x = 2uv, \qquad y = u^2 - v^2, \qquad z = u^2 + v^2$$

produces a Pythagorean triple x, y, z. In order for this triple to be primitive u and v must satisfy additional restrictions. It is easy to see from this that each integer $a > 2$ satisfies an equation of the form

$$a^2 + b^2 = z^2.$$

For if a is even, say $a = 2m$, then taking $u = m$ and $v = 1$ produces a triple with $a = x$. If a is odd, on the other hand, the reader should check that taking $u = (a + 1)/2$ and $v = (a - 1)/2$ produces a Pythagorean triple with $a = y$.

The situation is more complicated if a triple with a specified value of z is to be produced. Needed are integers u and v such that $u^2 + v^2 = z$. That there are infinitely many values of z for which this equation *cannot* be satisfied is a consequence of the following simple but important fact, which can be checked by squaring the elements in any complete residue system (mod 4).

THEOREM   If k is any integer, then

$$k^2 \equiv 0 \text{ or } 1 \quad (\text{mod } 4).$$

From this theorem we see that for any integers u and v we have

$$u^2 + v^2 \equiv 0, 1, \text{ or } 2 \quad (\text{mod } 4).$$

Thus no integer congruent to 3 (mod 4) can be written as the sum of two squares.

The Equation $n = u^2 + v^2$

We will investigate the general question of which integers n can be written as the sum of two squares. Being congruent to something other than 3 (mod 4) is not sufficient for an integer to be expressible this way. For example it is easily checked that

$$u^2 + v^2 = 6$$

has no solution.

For convenience we will allow u and v to be 0 or even negative. (Of course if u and v are solutions then so are $-u$ and $-v$.) For n up to 15 the representable numbers are $1 = 0^2 + 1^2$, $2 = 1^2 + 1^2$, $4 = 0^2 + 2^2$, $5 = 1^2 + 2^2$, $8 = 2^2 + 2^2$, $9 = 0^2 + 3^2$, $10 = 1^2 + 3^2$, and $13 = 2^2 + 3^2$.

Of great importance in the question of which integers are the sums of two squares is the identity

$$(a^2 + b^2)(c^2 + d^2) = (ac + bd)^2 + (ad - bc)^2, \tag{$*$}$$

which can be checked by multiplying out both sides.

LEMMA 1   If n_1 and n_2 can each be written as the sum of two squares, then so can $n_1 n_2$.

Proof Write $n_1 = a^2 + b^2$ and $n_2 = c^2 + d^2$ and apply (*).

Example Express 130 as the sum of two squares.

We know this can be done since we saw that 10 and 13 are so expressible. In fact taking $a = 1$, $b = 3$, $c = 2$, and $d = 3$ we see

$$130 = (ac + bd)^2 + (ad - bc)^2 = (2 + 9)^2 + (3 - 6)^2 = 11^2 + (-3)^2.$$

Lemma 1 suggests trying to determine which *primes* can be written as the sum of two squares. Up to 15 they are exactly 2, 5, and 13. Further investigation reveals that $17 = 1^2 + 4^2$ is the sum of two squares, but 19 is not; 23 is not, but $29 = 2^2 + 5^2$. The reader is encouraged to form his or her own conjecture before reading the next lemma.

LEMMA 2 The prime p can be written as the sum of two squares if and only if $p = 2$ or $p \equiv 1 \pmod 4$.

Proof We know $2 = 1^2 + 1^2$, and we also know that no number congruent to 3 (mod 4), prime or not, can be written as the sum of two squares. Thus it remains merely to show that any prime $p \equiv 1 \pmod 4$ is the sum of two squares. From part (e) of Theorem 5.3 we know that $(-1/p) = 1$; so there exists an integer x such that $x^2 \equiv -1 \pmod p$, or

$$x^2 + 1 \equiv 0 \pmod p. \tag{1}$$

The argument we now give makes use of the pigeon-hole principle. (See the problems at the end of Section 4.1.) Consider all the integers $rx + s$ with

$$0 \le r < \sqrt{p} \quad \text{and} \quad 0 \le s < \sqrt{p}.$$

Since r and s are allowed to be 0 there are more than $\sqrt{p}$ values of r and more than $\sqrt{p}$ values of s. Thus there are more than $\sqrt{p}\sqrt{p} = p$ pairs r, s. This means at least two of the numbers $rx + s$ must be congruent (mod p), say

$$r_1 x + s_1 \equiv r_2 x + s_2 \pmod p.$$

Thus we have

$$(r_1 - r_2)x \equiv s_2 - s_1 \pmod p. \tag{2}$$

If we let $u = |r_1 - r_2|$ and $v = |s_2 - s_1|$, then from (2) we have

$$ux \equiv \pm v \pmod p. \tag{3}$$

Notice that by the assumption that r_1, s_1 and r_2, s_2 are distinct pairs, u and v cannot both be 0.

Now from (1) and (3) we have

$$0 \equiv u^2(x^2 + 1) \equiv (ux)^2 + u^2 \equiv v^2 + u^2 \quad (\text{mod } p).$$

This means that $u^2 + v^2$ is some positive multiple of p. But

$$u^2 + v^2 < (\sqrt{p})^2 + (\sqrt{p})^2 = 2p,$$

so we must have $u^2 + v^2 = p$. We have proved that p is the sum of two squares.

The result that any prime congruent to 1 (mod 4) is the sum of two squares is the key to the analysis of $n = u^2 + v^2$. It was first stated by Albert Girard (1595–1632), and Fermat claimed to have a proof. (He probably did.) The first published proof, however, was by Euler in 1754.

THEOREM 7.1 A positive integer n can be written as the sum of two squares if and only if no prime congruent to 3 (mod 4) appears an odd number of times in the factorization of n into primes.

Proof First suppose no prime congruent to 3 (mod 4) appears an odd number of times in the factorization of n. Then we can write

$$n = k p_1 p_2 \cdots p_t,$$

where all primes congruent to 3 (mod 4) are collected into k, and each prime $p_1, p_2, \ldots, p_t$ is either 2 or congruent to 1 (mod 4). By hypothesis each prime in k appears to an even power, so $k = j^2$. That n is the sum of two squares now follows from repeated use of Lemma 1, since $k = 0^2 + j^2$ and each p_i is a sum of two squares from Lemma 2.

To prove the converse assume that $n = u^2 + v^2$. Let $d = (u, v)$, $U = u/d$, $V = v/d$, and $N = n/d^2$. Then U, V, and N are integers, and

$$N = U^2 + V^2. \tag{**}$$

Suppose some prime $p \equiv 3$ (mod 4) divides n to an odd power. Then p also divides N to an odd power, since dividing n by d^2 removes primes two at a time. If p divided U, then by (**) it would also divide V, contradicting the fact that $(U, V) = 1$. Thus there exists an integer x such that $xU \equiv 1$ (mod p). Multiplying (**) by x^2 and using the fact that p divides N, we get

$$1 + (xV)^2 \equiv 0 \quad (\text{mod } p).$$

Thus $(xV)^2 \equiv -1$ (mod p). But this contradicts Theorem 5.3(e), which says that -1 is not a quadratic residue modulo p for $p \equiv 3$ (mod 4).

Examples Earlier we found that among the integers from 1 to 15, exactly 1, 2, 4, 5, 8, 9, 10, and 13 could be represented as the sum of two squares. Now we can see why the remaining integers are not so representable. The integers 3, 6, 12, and 15 contain the factor 3 exactly once, 7 and 14 contain 7 exactly once, and 11 is itself congruent to 3 (mod 4).

The integers from 51 to 60 can be analyzed in a similar manner.

$51 = 3 \cdot 17$ (note the 3)

$52 = 2^2 13 = 4^2 + 6^2$

$53 = 2^2 + 7^2$

$54 = 2 \cdot 3^3$ (note the 3^3)

$55 = 5 \cdot 11$ (note the 11)

$56 = 2^3 7$ (note the 7)

$57 = 3 \cdot 19$ (note 3 and 19)

$58 = 2 \cdot 29 = 3^2 + 7^2$

$59 \equiv 3 \pmod 4$

$60 = 2^2 3 \cdot 5$ (note the 3)

PROBLEMS FOR SECTION 7.2

In the first eight problems tell whether the given integer can be written as the sum of two squares.

1. 61	2. 62	3. 63	4. 65
5. 68	6. 117	7. $51 \cdot 45$	8. $39 \cdot 54$

In the next six problems use the identity (∗) of this section to find positive integers u and v such that $u^2 + v^2$ equals the given integer.

9. $10 \cdot 17$	10. $18 \cdot 5$	11. 85
12. 221	13. 261	14. 637

15. Which integers from 1 to 10 can be written as the sum of three squares?

16. Which integers from 11 to 20 can be written as the sum of three squares?

17. Prove that $a^2 \equiv 0, 1,$ or $4 \pmod 8$ for all integers a.

18. Prove that if $n \equiv 7 \pmod 8$, then n cannot be expressed as the sum of three squares.

19. Show that if $4 \mid x^2 + y^2 + z^2$, then x, y, and z are all even.

20. Show that no integer of the form $4^k(8m + 7)$ can be written as the sum of three squares. (**Hint:** Use the two previous problems.)

21. Show that if p is a prime congruent to 1 (mod 4), then there exist u and v such that $p = u^2 + v^2$ with $0 < u < v$ and $(u, v) = 1$.

22. Show that if p and q are distinct primes congruent to 1 (mod 4), then $pq = u^2 + v^2$, $0 < u < v$, has two distinct solutions. (**Hint:** Use the previous problem and the identity (∗).)

23. Show that if $n \equiv 2$ (mod 4), then n cannot be written as the difference of two squares.

24. Show that any positive integer $n \not\equiv 2$ (mod 4) can be written as the difference of two squares. (**Hint:** See the discussion at the beginning of this section for the case when n is odd.)

In the next four problems assume that $p = a^2 + b^2 = c^2 + d^2$, *where* $0 < a < b$, $0 < c < d$, *and* p *is prime.*

25. Show that $a^2d^2 \equiv b^2c^2$ (mod p).

26. Show that $ad - bc = 0$ or $ad + bc = p$. (**Hint:** Each of a, b, c, and d is $< \sqrt{p}$.)

27. Show that $ad = bc$. (**Hint:** Use (∗).)

28. Show that $a = c$ and $b = d$, so that the representation of a prime as the sum of two squares is essentially unique. (**Hint:** Show that $c = ka$, and that $k = 1$.)

7.3 Sums of Four Squares

In the last section we saw that not every integer can be written as the sum of two squares. Even three squares does not suffice to represent some numbers, for example 7. In the problems at the end of Section 7.2 it is shown that no integer of the form $4^k(8m + 7)$ can be written as the sum of three squares. Actually every positive integer not of this form can be expressed as the sum of three squares, although we will not prove that result. One reason that the analysis of sums of three squares is difficult is that there is no identity similar to the one labeled (∗) in the last section. In fact it is not true that the product of numbers that are the sum of three squares is also of that form, as is shown by the example

$$2 \cdot 14 = (0^2 + 1^2 + 1^2)(1^2 + 2^2 + 3^2) = 28 = 4(8 \cdot 0 + 7).$$

It turns out that if four squares are allowed, then every positive integer can be represented. This result, which we will prove in this section, may have been known to Diophantus, although he does not state it explicitly in any of his known writings. Fermat claimed to have a proof of it, but, as was his custom, he published none. Euler worked on the problem for many years, but never found a proof. The first published proof was by Lagrange in 1772, using many of Euler's partial results. Among these was the identity showing that the product of two numbers, each the sum of four squares, is also of that form. We state that identity now; the reader should prove it by multiplying out both sides.

LEMMA 1 $(a^2 + b^2 + c^2 + d^2)(A^2 + B^2 + C^2 + D^2) = (aA + bB + cC + dD)^2 + (aB - bA + cD - dC)^2 + (aC - bD - cA + dB)^2 + (aD + bC - cB - dA)^2$.

Because of Lemma 1 it suffices to prove that each prime can be written as the sum of four squares. But Theorem 7.1 says that all primes except those congruent to 3 (mod 4) can be written as the sum of just two squares, so we need only worry about primes of the form $4m + 3$.

LEMMA 2   Let p be a prime congruent to 3 (mod 4). Then there exists an integer k with $0 < k < p$ such that kp is the sum of four squares.

Proof   Let t be the smallest positive integer that is not a quadratic residue (mod p). Clearly $t > 1$. Then by parts (b) and (e) of Theorem 5.3 we have

$$\left(\frac{-t}{p}\right) = \left(\frac{-1}{p}\right)\left(\frac{t}{p}\right) = (-1)(-1) = 1.$$

This means that $-t$ is a quadratic residue (mod p). Thus there exists an integer x with $x^2 \equiv -t$ (mod p), with $0 < x < p/2$. (See the theorem in Section 5.3 following the definition of quadratic residue.)

Now by the definition of t the positive integer $t - 1$ must be a quadratic residue (mod p), and so in the same way there exists y such that $y^2 \equiv t - 1$ (mod p) and $0 < y < p/2$. Then

$$x^2 + y^2 + 1 \equiv -t + (t - 1) + 1 \equiv 0 \quad \text{(mod } p),$$

so $x^2 + y^2 + 1 = kp$ for some positive integer k. But then

$$kp = x^2 + y^2 + 1 < \left(\frac{p}{2}\right)^2 + \left(\frac{p}{2}\right)^2 + 1 = \frac{p^2}{2} + 1 < \frac{p^2}{2} + \frac{p^2}{2} = p^2,$$

and so $k < p$. Since $kp = x^2 + y^2 + 1^2 + 0^2$, the Lemma is proved.

Example   We will illustrate this lemma with $p = 7$. The quadratic residues are $1^2 = 1$, $2^2 = 4$, and $3^2 \equiv 2$ (mod 7). Thus the smallest positive nonresidue is $t = 3$. Since $-t = -3 \equiv 4 \equiv 2^2$ (mod 7) we take $x = 2$. Likewise $t - 1 = 2 \equiv 3^2$ (mod 7), so $y = 3$. Then $x^2 + y^2 + 1 = 4 + 9 + 1 = 14 = 2 \cdot 7$, so $k = 2 < p$.

THEOREM (Lagrange)   Every positive integer can be written as the sum of four squares.

Proof We have seen that it suffices to prove that any prime congruent to 3 (mod 4) is the sum of four squares. Let p be such a prime. From Lemma 2 there exists a positive integer $k < p$ such that kp is the sum of four squares. (Actually the proof from this point on will work for any odd prime p such that such a k exists.) Of all such integers k let m be the smallest, so we have

$$mp = a^2 + b^2 + c^2 + d^2, \quad 0 < m < p. \tag{1}$$

First we show m is odd. For if m is even, then of the integers a, b, c, and d exactly none, two, or four must be even. If two of them are even, relabel them so that they are a and b. Then in any case $a - b$, $a + b$, $c - d$, and $c + d$ are all even. But then

$$\left(\frac{a-b}{2}\right)^2 + \left(\frac{a+b}{2}\right)^2 + \left(\frac{c-d}{2}\right)^2 + \left(\frac{c+d}{2}\right)^2$$
$$= \frac{a^2 + b^2 + c^2 + d^2}{2} = \left(\frac{m}{2}\right)p,$$

which contradicts the minimality of m.

Now if $m = 1$, we are done, for then p is the sum of four squares. Assume $m > 1$. Since the m consecutive integers

$$-\frac{m-1}{2}, \quad -\frac{m-1}{2} + 1, \ldots, 0, 1, \ldots, \frac{m-1}{2}$$

form a complete residue system (mod m) we can choose elements A, B, C, and D from among them (so each is less than $m/2$ in absolute value) such that

$$A \equiv a, \quad B \equiv b, \quad C \equiv c, \quad \text{and} \quad D \equiv d \pmod{m}.$$

Then

$$A^2 + B^2 + C^2 + D^2 \equiv a^2 + b^2 + c^2 + d^2 = mp \equiv 0 \pmod{m}.$$

Let

$$A^2 + B^2 + C^2 + D^2 = rm. \tag{2}$$

Note that $r > 0$, for if A, B, C, and D are all 0, then m divides each of a, b, c, and d. But then m^2 divides $a^2 + b^2 + c^2 + d^2 = mp$, which is impossible since $1 < m < p$. Notice also that

$$rm = A^2 + B^2 + C^2 + D^2 < 4(m/2)^2 = m^2,$$

and so $r < m$.

Now we will make use of the actual identity of Lemma 1. From (1) and (2) we have

$$(mp)(rm) = (a^2 + b^2 + c^2 + d^2)(A^2 + B^2 + C^2 + D^2) = w^2 + x^2 + y^2 + z^2,$$

where

$$w = aA + bB + cC + dD \equiv a^2 + b^2 + c^2 + d^2 = mp \equiv 0 \quad (\text{mod } m),$$

$$x = aB - bA + cD - dC \equiv ab - ba + cd - dc \equiv 0 \quad (\text{mod } m),$$

$$y = aC - bD - cA + dB \equiv ac - bd - ca + db \equiv 0 \quad (\text{mod } m),$$

$$z = aD + bC - cB - dA \equiv ad + bc - cb - da \equiv 0 \quad (\text{mod } m).$$

But then

$$rp = (w/m)^2 + (x/m)^2 + (y/m)^2 + (z/m)^2,$$

which contradicts the minimality of m, since $0 < r < m$.

Although it is natural to use Theorem 7.1 to reduce the problem to primes congruent to 3 (mod 4), the four-square theorem can also be proved without any knowledge of sums of two squares. All we need do is include in Lemma 2 primes congruent to 1 (mod 4) also. It is also possible to prove Lemma 2 (for all odd primes) without using any results on quadratic residues. These proofs are worked out in the exercises.

PROBLEMS FOR SECTION 7.3

 In the first six problems find all quadruples a, b, c, d with $0 \leq a \leq b \leq c \leq d$ and $n = a^2 + b^2 + c^2 + d^2$.

1. $n = 20$ 2. $n = 27$ 3. $n = 30$
4. $n = 31$ 5. $n = 32$ 6. $n = 36$

In the next four problems work through the proof of Lemma 2 for the given value of p. What x, y, and k result?

7. $p = 11$ 8. $p = 19$ 9. $p = 23$ 10. $p = 31$

In the next four problems use $7 = 1^2 + 1^2 + 1^2 + 2^2$, $11 = 0^2 + 1^2 + 1^2 + 3^2$, $10 = 1^2 + 1^2 + 2^2 + 2^2$, and Lemma 1 to express each integer as the sum of four squares.

11. $7 \cdot 11$ 12. $11 \cdot 10$ 13. $7 \cdot 10$ 14. $77 \cdot 10$

In each of the next four problems we have mp written as the sum of four squares, where $1 < m < p$. Use the methods of the proof of the theorem of this section to write a smaller multiple of p as a sum of four squares. (The method will depend on whether m is even or odd.)

15. $p = 19$, $6p = 8^2 + 7^2 + 1^2 + 0^2$ 16. $p = 29$, $4p = 9^2 + 5^2 + 3^2 + 1^2$

17. $p = 11$, $3p = 5^2 + 2^2 + 2^2 + 0^2$ 18. $p = 17$, $5p = 8^2 + 4^2 + 2^2 + 1^2$

B

19. In Lemma 2 it is proved that there exist nonnegative integers x and y, each less than $p/2$, such that $x^2 + y^2 + 1 \equiv 0 \pmod{p}$. Show that this holds also for primes $p \equiv 1 \pmod 4$ without using Theorem 7.1. (**Hint:** $(-1/p) = 1$.)

20. Which integers from 1 to 20 cannot be expressed as the sum of four *positive* squares?

21. Which integers from 21 to 40 cannot be expressed as the sum of four positive squares?

22. Show that 169 can be written as the sum of one, two, three, and four positive squares.

23. Show that if $n > 169$, then n can be written as the sum of five positive squares. (**Hint:** Write $n - 169$ as the sum of four squares, some of which may be 0.)

24. Show that if $8 | a^2 + b^2 + c^2 + d^2$, then a, b, c, and d are all even.

25. Show by induction that $4^n 2$ is not the sum of four positive squares for all positive integers n.

C

The following three problems indicate a proof of Lemma 2 that does not use Theorem 7.1 or quadratic residue theory, and that applies to all odd primes. Assume that p is an odd prime.

26. Let $S = \{0^2, 1^2, 2^2, \ldots, ((p - 1)/2)^2\}$. Show that no two elements of S are congruent $\pmod p$.

27. Let $T = \{-s - 1 : s \text{ in } S\}$, where S is defined in the previous problem. Show that no two elements of T are congruent $\pmod p$, but that some element x^2 of S is congruent to some element $-y^2 - 1$ of T modulo p, where x and y are nonnegative and less than $p/2$. (**Hint:** The pigeon-hole principle.)

28. Show that $x^2 + y^2 + 1 = mp$, with $0 < m < p$, where x and y are as in the previous problem.

7.4 Sums of Fourth Powers

Waring's Problem for Fourth Powers

In Section 4.4 we mentioned Waring's Problem, which is to show that for each $k > 1$ there is a minimal integer $g(k)$ such that each positive integer can be written as a sum of $g(k)$ nonnegative kth powers. The 1772 theorem of Lagrange proved in the last section (along with the fact that 7 cannot be written as the sum of 3 squares) shows that $g(2) = 4$. Further progress on Waring's Problem was not made until 1859, when Liouville showed that $g(4) \le 53$, which is a compact way of stating the next theorem.

THEOREM Every positive integer can be written as the sum of 53 fourth powers.

Proof We need the identity

$$6(a^2 + b^2 + c^2 + d^2)^2 = (a + b)^4 + (a - b)^4 + (c + d)^4 + (c - d)^4$$
$$+ (a + c)^4 + (a - c)^4 + (b + d)^4 + (b - d)^4 \quad (*)$$
$$+ (a + d)^4 + (a - d)^4 + (b + c)^4 + (b - c)^4.$$

Although this can be verified by multiplying out both sides, we offer a little help in cutting down the work. The left side is

$$6(a^4 + b^4 + c^4 + d^4) + 12(a^2b^2 + a^2c^2 + \cdots + c^2d^2).$$

Since each variable appears in six terms on the right side of $(*)$, and since

$$(x + y)^4 = x^4 + 4x^3y + 6x^2y^2 + 4xy^3 + y^4,$$

the fourth powers match up properly. Likewise each pair of variables (such as a and b) appears in two terms on the right side of $(*)$, accounting for 12 times the product of their squares. It can be seen that the terms involving odd exponents cancel out, since each pair of variables appears once with a plus sign and once with a minus sign.

Now by Lagrange's theorem each positive integer is the sum of four squares, say $a^2 + b^2 + c^2 + d^2$. Thus $(*)$ implies that each number that is six times a square can be written as the sum of 12 fourth powers.

Let n be an arbitrary positive integer. Write

$$n = 6q + r, \quad 0 \le r < 6.$$

Then if we write $q = w^2 + x^2 + y^2 + z^2$ (using Lagrange's theorem again), we have

$$n = 6w^2 + 6x^2 + 6y^2 + 6z^2 + r.$$

Each of the first four terms in this expression can be written as the sum of 12 fourth powers, as we saw above, and r, being at most 5, can be written as the sum of 5 fourth powers (of 0 and 1). Thus n is the sum of $12 + 12 + 12 + 12 + 5 = 53$ fourth powers.

By being a little more careful it is not hard to improve the proof just given to show that $g(4) \le 50$; the details are given in the exercises. Actually it is known that $19 \leq g(4) \leq 35$. That $g(4) \geq 19$ follows from the fact that 79 cannot be written as the sum of 18 fourth powers. For $3^4 = 81$ is already too big, so only the fourth powers 1 and 16 are available. Using four 16's still requires $79 - 4(16) = 15$ ones, a total of 19 fourth powers.

Fermat's Last Theorem

Fermat's Last Theorem (which, despite its name, has never been proved in general) was introduced in Section 2.3. It says that for $k > 2$ the equation

$$x^k + y^k = z^k \tag{1}$$

has no solutions in positive integers. We will prove the case $k = 4$. In fact a somewhat stronger result will be proved, namely that

$$x^4 + y^4 = w^2 \tag{2}$$

has no solutions in positive integers. This implies the $k = 4$ case of Fermat's Last Theorem since if (1) had a solution with $k = 4$, then setting $w = z^2$ would give a solution to (2). The proof will depend on our analysis of Pythagorean triples in Section 7.1.

THEOREM   The equation $x^4 + y^4 = w^2$ has no solution in positive integers.

Proof   We will start by showing that if equation (2) has a solution in positive integers x, y, and w, then it has another solution X, Y, W in positive integers with $W < w$.

Case 1:   Some prime p divides all of x, y, and w.
Then p^4 divides $x^4 + y^4 = w^2$, from which we see that p^2 divides w. Thus

$$\left(\frac{x}{p}\right)^4 + \left(\frac{y}{p}\right)^4 = \left(\frac{w}{p^2}\right)^2,$$

and so $X = x/p$, $Y = y/p$, $W = w/p^2$ is another solution with $W < w$.

Case 2:   No prime divides all of x, y, and w.
Then since

$$(x^2)^2 + (y^2)^2 = w^2, \tag{3}$$

x^2, y^2, and w form a primitive Pythagorean triple. By the lemma of Section 7.1 we conclude that x^2, y^2, and w (and so x, y, and w) are relatively prime in pairs, that exactly one of x^2 and y^2 is even, and that w is odd. Let us assume that x^2 is even, so that x is even and y is odd.

Then by the Pythagorean Triple Theorem applied to (3) there exist relatively prime positive integers u and v such that

$$x^2 = 2uv, \qquad y^2 = u^2 - v^2, \qquad \text{and} \qquad w = u^2 + v^2.$$

We now apply the same theorem to the equation

$$y^2 + v^2 = u^2.$$

We see that y, v, and u form another primitive Pythagorean triple, and since we know that y is odd we must have v even and u odd. Thus there exist relatively prime positive integers r and s such that

$$y = r^2 - s^2, \qquad v = 2rs, \qquad \text{and} \qquad u = r^2 + s^2.$$

Now since $(u, v/2) = 1$, by considering the prime factorization of both sides of the equation $(x/2)^2 = u(v/2)$ we see that there must exist positive integers W and Z such that $u = W^2$ and $v/2 = Z^2$. In the same way since $(r, s) = 1$ from the equation $rs = v/2 = Z^2$ we see that there exist positive integers X and Y such that $r = X^2$ and $s = Y^2$.

Now we have

$$X^4 + Y^4 = r^2 + s^2 = u = W^2, \tag{4}$$

where X, Y, and W are positive integers. Also

$$w = u^2 + v^2 = W^4 + v^2 > W^4 \geq W.$$

From equation (4) we see we have another solution to equation (2) with $W < w$, as claimed at the beginning of this proof.

So far our argument has been entirely positive. We showed that given a solution x, y, w to (2) in positive integers, then we can find another solution X, Y, W, also in positive integers, with $W < w$. But we need not stop here. From X, Y, and W we could create a third solution, say X', Y', W', with $W' < W$. In fact, given just one solution, we could find an infinite sequence of solutions, with

$$w > W > W' > W'' > \cdots > 0.$$

This is clearly impossible, since one cannot have an infinite decreasing sequence of positive integers. Thus no solution can exist.

The method of proof just employed is due to Fermat, and is known as the "method of infinite descent." The argument could also be arranged as an indirect proof depending on the principle that any nonempty set of positive integers has a smallest element. Assuming a solution in positive integers to (2) exists, we let x, y, w be that positive solution with w minimal. Then showing the existence of the solution X, Y, W with $W < w$ provides a contradiction. Since the principle that any nonempty set of positive integers has a least element (called the "Well-Ordering Principle") can be shown to be equivalent to Mathematical Induction, infinite descent can be considered to be a variation on induction. See the problems at the end of Sections 1.6 and 2.1 for the equivalence of the Mathematical Induction and Well-Ordering Principles.

PROBLEMS FOR SECTION 7.4

In the first six problems write n as the sum of as few positive cubes as possible.

1. $n = 20$ 2. $n = 30$ 3. $n = 40$
4. $n = 52$ 5. $n = 100$ 6. $n = 155$

In the next four problems write n as the sum of as few positive fourth powers as possible.

7. $n = 100$ 8. $n = 177$ 9. $n = 200$ 10. $n = 300$

11. Find the smallest positive integer that cannot be written as the sum of eight nonnegative cubes.
12. Find the second smallest positive integer that cannot be written as the sum of 18 fourth powers.

13. Show that if $0 < n < 79$, then n can be written as the sum of 18 fourth powers.
14. Show that if $n > 80$, then $n = 6q + r$, where $r = 0, 1, 2, 16, 17,$ or 81.
15. Show that $g(4) \leq 50$.
16. Let k be a positive integer, and define $t = [(3/2)^k]$. Show that $2^k t - 1$ cannot be expressed as the sum of fewer than $2^k + t - 2$ nonnegative kth powers. (**Hint:** Show $2^k t - 1 < 3^k$.)
17. Show that $x^4 \equiv 0$ or $1 \pmod{16}$ for all integers x. Conclude that no number congruent to 15 (mod 16) can be written as the sum of fewer than 15 fourth powers.
18. Show that if $n \equiv 4$ or $5 \pmod 9$ then n cannot be written as the sum of three nonnegative cubes.
19. Show that if n is any positive integer then $x^{4n} + y^{4n} = z^{4n}$ is impossible in positive integers.
20. Show that to prove Fermat's Last Theorem it suffices to show that if p is any odd prime then $x^p + y^p = z^p$ is impossible in positive integers.
21. Show that $x^4 - 4y^4 = w^2$ has no solutions in positive integers. (**Hint:** Compute $w^4 + (2xy)^4$.)

22. Show that given any positive integer n there exists a positive integer a such that $n + a^2 < (a + 1)^2$.

C 23. One way to represent a positive integer n as a sum of squares is to let $x_1 = [\sqrt{n}]$, so $x_1^2 \le n$. Then let $n_1 = n - x_1^2$. If $n_1 > 0$ let $x_2 = [\sqrt{n_1}]$, and $n_2 = n_1 - x_2^2$. Eventually we get $n_{k+1} = 0$, and $n = x_1^2 + x_2^2 + \cdots + x_k^2$. (This is called the "greedy algorithm," since at each step we take as big a bite out of n as possible.) Show that given k there exists n such that this algorithm represents n as the sum of k nonzero squares. (**Hint:** See the previous exercise.)

In the following four problems assume $x^4 - y^4 = w^2$ has a solution in positive integers, and that x, y, w is such a solution with x minimal.

24. Show that x, y, and w are relatively prime in pairs.
25. Show that x is odd.
26. Show that if y is odd there exist relatively prime positive integers u and v with $(xy)^2 = u^4 - v^4$ and $u < x$.
27. Show that if y is even there exist relatively prime positive integers u and v with v even such that $y^2 = 2uv$ and $x^2 = u^2 + v^2$. Then show there exist relatively prime positive integers a and b such that $u = a^2$, $v/2 = b^2$, and $x^2 = a^4 + 4b^4$. Show then there exist relatively prime positive integers U and V such that $a^2 = U^2 - V^2$ and $b^2 = UV$. Then show there exist positive integers c and d such that $U = c^2$, $V = d^2$, $a^2 = c^4 - d^4$, and $c < x$.

28. Show that the equation $x^4 - y^4 = w^2$ has no solutions in positive integers.

7.5   Pell's Equation

A Misnamed Diophantine Equation

The equation $x^2 - dy^2 = c$, where d and c are given constants and x and y are the unknowns, is known as **Pell's equation**, due to a mistake by Euler. The title is unfortunate, since the English mathematician John Pell (1610–1685) made no contribution to the theory of the equation. If $d < 0$, clearly x and y cannot be too large; and if solutions exist, there are only a finite number, which can be found by trial and error.

Likewise if d is a perfect square, say $d = a^2$, then the equation can be written as

$$(x + ay)(x - ay) = c,$$

and all solutions may be found by considering the possible factorizations of c.

Example Find all nonnegative solutions to $x^2 - 9y^2 = -161$.

Since $161 = 7 \cdot 23$, and since $x + 3y$ cannot be negative, the possibilities for $x + 3y$ and $x - 3y$ are 161 and -1, 1 and -161, 23 and -7, and 7 and -23. Only the first and third cases lead to nonnegative integral values for x and y, namely, $x = 80$ and $y = 27$, and $x = 8$ and $y = 5$.

From now on we will concentrate on the case of Pell's equation when infinitely many solutions are at least a possibility, namely when $d > 0$ and d is not a perfect square. Then $\sqrt{d}$ is irrational by the first theorem of Section 6.1. Clearly it suffices to look for nonnegative solutions; the other solutions can easily be generated from these. We will usually consider cases when c is small compared to d, and our strongest result will be for $c = 1$. Then x^2 will have to be close to dy^2 for x and y to be a solution, and so, dividing by y^2, $(x/y)^2$ will be close to d. This means that x/y will be close to the irrational number $\sqrt{d}$.

In the last chapter we mentioned that the convergents to an irrational number comprise good approximations to that number. The next theorem makes that statement somewhat more definite.

THEOREM Suppose $d > 0$ is not a perfect square, and let $[q_1, q_2, \ldots]$ be the continued fraction expansion for $\sqrt{d}$. Let $h_i/k_i = [q_1, \ldots, q_i]$ be the convergents to $\sqrt{d}$, where the integers h_i and k_i are as defined in Section 6.3, $i = -1, 0, \ldots$. Then

$$|h_i^2 - dk_i^2| < 2\sqrt{d} + 1$$

for all $i \geq 0$.

Proof The inequality is clear for $i = 0$ since $h_0 = 1$ and $k_0 = 0$. By Theorems 6.3 and 6.4 of Section 6.3 we have

$$\frac{h_1}{k_1} < \frac{h_3}{k_3} < \cdots < \sqrt{d} < \cdots < \frac{h_4}{k_4} < \frac{h_2}{k_2}.$$

(See (∗∗) in that section.) Thus for any $i > 1$ we have

$$\left|\frac{h_{i-1}}{k_{i-1}} - \sqrt{d}\right| < \left|\frac{h_{i-1}}{k_{i-1}} - \frac{h_i}{k_i}\right|.$$

By (∗) of Section 6.3 the latter quantity equals $1/k_i k_{i-1}$. Then $|h_{i-1} - k_{i-1}\sqrt{d}| < 1/k_i$, and so $h_{i-1} = k_{i-1}\sqrt{d} + E$, where $|E| < 1/k_i$. Thus

$$h_{i-1}^2 = k_{i-1}^2 d + 2k_{i-1}E\sqrt{d} + E^2,$$

and

$$|h_{i-1}^2 - k_{i-1}^2 d| = |2k_{i-1}\sqrt{d}\,E + E^2| < \frac{2k_{i-1}\sqrt{d}}{k_i} + \frac{1}{k_i^2} \leq 2\sqrt{d} + 1,$$

since the numbers k_i form a nondecreasing sequence of positive integers for $i > 0$.

Example If $d = 3$ we compute $\sqrt{3} = [1, 1, 2, 1, 2, \ldots] = [1, \overline{1, 2}]$ by the methods of Section 6.3. Notice that all the values of $h_i^2 - 3k_i^2$ below are less in absolute value than $2\sqrt{3} + 1$, which is about 4.46.

i	h_i	k_i	q_i	$h_i^2 - 3k_i^2$
-1	0	1		
0	1	0		
1	1	1	1	$1 - 3 = -2$
2	1	1	1	$4 - 3 = 1$
3	5	3	2	$25 - 3(9) = -2$
4	7	4	1	$49 - 3(16) = 1$
5	19	11	2	$361 - 3(121) = -2$

Now we consider the equation $x^2 - dy^2 = 1$, where $d > 0$ is not a perfect square. Clearly x cannot be 0; and if $y = 0$ then x must be 1 or -1. Thus we will concentrate on solutions where both x and y are positive.

THEOREM If $d > 0$ is not a perfect square, then $x^2 - dy^2 = 1$ has a solution in positive integers x and y.

Proof By the previous theorem there are infinitely many pairs of relatively prime positive integers x and y such that $|x^2 - dy^2| < 2\sqrt{d} + 1$. Thus there must exist an integer $t \neq 0$ such that $x^2 - dy^2 = t$ for infinitely many such x and y. (We cannot have $x^2 - dy^2 = 0$ for positive x and y since then we would have $\sqrt{d} = x/y$, a rational number.)

If r is the least residue of $x \pmod{|t|}$ and s is the least residue of $y \pmod{|t|}$, then the pair (r, s) can assume only t^2 values. Thus there exist two pairs x, y and X, Y such that

$$x^2 - dy^2 = t, \qquad X^2 - dY^2 = t,$$

$$x \equiv X \pmod{|t|}, \quad \text{and} \quad y \equiv Y \pmod{|t|}.$$

Note that

$$xX - yYd \equiv x^2 - y^2d = t \equiv 0 \pmod{|t|},$$

and $\quad xY - Xy \equiv xy - xy = 0 \pmod{|t|},$

so there exist integers u and v such that

$$xX - yYd = tu \quad \text{and} \quad xY - Xy = tv.$$

Then

$$t^2 = (x^2 - dy^2)(X^2 - dY^2)$$

$$= (x - y\sqrt{d})(x + y\sqrt{d})(X - Y\sqrt{d})(X + Y\sqrt{d})$$

$$= (x - y\sqrt{d})(X + Y\sqrt{d})(x + y\sqrt{d})(X - Y\sqrt{d})$$

$$= (xX - yYd + (xY - Xy)\sqrt{d})(xX - yYd + (-xY + Xy)\sqrt{d})$$

$$= (tu + tv\sqrt{d})(tu - tv\sqrt{d}) = t^2(u^2 - dv^2),$$

and so $u^2 - dv^2 = 1$.

It remains to show that $v \neq 0$. By its definition if $v = 0$, then $xY = Xy$, and $x/y = X/Y$. But we know the convergents to an irrational number are distinct.

Infinitely Many Solutions

Although we will not prove it here, a solution to $x^2 - dy^2 = 1$ will always be found among the convergents to the continued fraction expansion of $\sqrt{d}$. This is important because the next theorem shows how to generate infinitely many solutions to this equation from a single one, and then how to generate infinitely many solutions to the more general equation

$$x^2 - dy^2 = c,$$

given one solution.

THEOREM Suppose $d > 0$ is not a perfect square.

  a. If x, y is any positive solution to

$$x^2 - dy^2 = 1, \tag{1}$$

  and if $n > 0$, then the integers x_n, y_n uniquely defined by $x_n + y_n\sqrt{d} = (x + y\sqrt{d})^n$ are also a positive solution to (1), and these are distinct for distinct values of n.

  b. If u, v is any positive solution to

$$u^2 - dv^2 = c, \tag{2}$$

  and if x, y is any of the infinitely-many positive solutions to (1), then the pair U, V defined by

$$U = ux + vyd, \qquad V = uy + vx$$

is also a positive solution to (2), and these are distinct for distinct positive solutions x, y.

Proof (a) The first question is whether x_n and y_n are uniquely defined. Expanding $(x + y\sqrt{d})^n$ gives an expression of the form $X + Y\sqrt{d}$, where X and Y are positive integers. Suppose $X + Y\sqrt{d} = U + V\sqrt{d}$ for any integers X, Y, U, and V. Then $(Y - V)\sqrt{d} = U - X$. If $Y - V$ is not 0, then this equation implies that $\sqrt{d}$ is rational. Thus $V = Y$ and so $U = X$.

Notice that if we expand instead $(x - y\sqrt{d})^n$ all the odd powers of $\sqrt{d}$ enter with a minus sign and we get $X - Y\sqrt{d} = x_n - y_n\sqrt{d}$. Thus

$$x_n^2 - dy_n^2 = (x_n - y_n\sqrt{d})(x_n + y_n\sqrt{d})$$
$$= (x - y\sqrt{d})^n(x + y\sqrt{d})^n = (x^2 - dy^2)^n = 1^n = 1.$$

We see that x_n, y_n is a solution for each n.

Since $x + y\sqrt{d} > 1$ the numbers $x_n + y_n\sqrt{d} = (x + y\sqrt{d})^n$ form a strictly increasing sequence. Thus the pairs x_n, y_n must be distinct. In fact, since $x_n^2 - dy_n^2 = 1$, both sequences x_n and y_n increase.

(b) From equations (1) and (2) we have

$$U^2 - dV^2 = u^2x^2 + 2uvxyd + v^2y^2d^2 - d(u^2y^2 + 2uvxy + v^2x^2)$$
$$= u^2(x^2 - dy^2) - dv^2(x^2 - dy^2) = u^2 - dv^2 = c.$$

If x, y and x', y' are two solutions to (1) with $x > x'$, then $y > y'$ also. Then $U' > U$ and $V' > V$, where U' and V' are defined in the obvious way.

Example　Find two more positive solutions to $x^2 - 7y^2 = 1$ besides $x = 8$ and $y = 3$.

We have $(8 + 3\sqrt{7})^2 = 127 + 48\sqrt{7}$, so $x = 127$ and $y = 48$ is another solution. Likewise $(8 + 3\sqrt{7})^3 = (8 + 3\sqrt{7})(127 + 48\sqrt{7}) = 2024 + 765\sqrt{7}$, so $x = 2024$ and $y = 765$ is a third solution.

Example　Find another positive solution to $x^2 - 7y^2 = -3$ besides $x = 5$ and $y = 2$.

We take $u = 5$, $v = 2$, and, from the previous example, $x = 8$ and $y = 3$. Then $U = 5 \cdot 8 + 2 \cdot 3 \cdot 7 = 82$, $V = 5 \cdot 3 + 2 \cdot 8 = 31$ satisfies $U^2 - 7V^2 = -3$.

The value $c = 1$ is the only one for which Pell's equation is always assured of having a solution. Although we used continued fractions to show the existence of solutions in this case, we have not indicated the full connection between solutions of $x^2 - dy^2 = c$ and the continued fraction expansion of $\sqrt{d}$. For example, it can be shown that if $|c| < \sqrt{d}$, then any positive solution to this equation corresponds to the numerator and denominator of a convergent h_i/k_i of the expansion. Since the quadratic irrational $\sqrt{d}$ has a periodic continued fraction expansion, the question of for which values of c with $|c| < \sqrt{d}$ the equation has solutions, and the form of those solutions, is reduced to a finite determination. We have not gone into the theory of the continued fraction expansions of quadratic irrationalities sufficiently to give details. The reader is referred to the book by Niven and Zuckerman listed in the references for a fuller treatment.

DEFINITION   minimal solution

Suppose $d > 0$ is not a perfect square. We say that x and y satisfying $X^2 - dY^2 = c$ is a **minimal solution** if x and y are positive, and if whenever x', y' is any other positive solution, then $x' > x$. There can be only one minimal solution. It turns out that if x, y is the minimal solution to $X^2 - dY^2 = 1$, then all positive solutions are generated as in part (a) of the last theorem. Also if x, y is the minimal solution to $X^2 - dY^2 = -1$, then u, v is the minimal solution to $X^2 - dY^2 = 1$, where $u + v\sqrt{d} = (x + y\sqrt{d})^2$. Proofs of these results will be found in the exercises.

PROBLEMS FOR SECTION 7.5

 In the first six problems use part (a) of the last theorem of this section to find two more positive solutions to the equation $X^2 - dY^2 = 1$, given that the x and y form a solution.

1. $d = 2, x = 3, y = 2$ 2. $d = 15, x = 4, y = 1$
3. $d = 3, x = 5, y = 2$ 4. $d = 12, x = 7, y = 4$
5. $d = 11, x = 10, y = 3$ 6. $d = 63, x = 8, y = 1$

In the next four problems compute $h_i^2 - dk_i^2$ for $i = 1, 2, 3,$ and 4, where h_i/k_i is the ith convergent to $\sqrt{d}$.

7. $d = 7$ 8. $d = 13$ 9. $d = 12$ 10. $d = 10$

In the next four problems find the smallest two solutions to the given equation where x/y is a convergent to $\sqrt{d}$.

11. $x^2 - 8y^2 = -4$ ($\sqrt{8} = [2, \overline{1, 4}]$)
12. $x^2 - 6y^2 = 1$ ($\sqrt{6} = [2, \overline{2, 4}]$)
13. $x^2 - 5y^2 = 1$ ($\sqrt{5} = [2, \overline{4}]$)
14. $x^2 - 11y^2 = -2$ ($\sqrt{11} = [3, \overline{3, 6}]$)

In the next four problems find by inspection the minimal solution to $x^2 - dy^2 = 1$ and $x^2 - dy^2 = c$, then use the method of part (b) of the last theorem of this section to find another solution to $x^2 - dy^2 = c$.

15. $d = 24, c = -20$ 16. $d = 6, c = 3$
17. $d = 5, c = -19$ 18. $d = 30, c = 19$

In the next six problems find all nonnegative solutions to the given equations.

19. $x^2 + 3y^2 = 28$ 20. $x^2 + 5y^2 = 69$ 21. $x^2 - 4y^2 = 45$
22. $x^2 - 9y^2 = 64$ 23. $x^2 - 16y^2 = -23$ 24. $x^2 - 25y^2 = -56$

B

In the remaining problems assume that $d > 0$ is not a perfect square.

25. Let X, Y be a positive solution to $x^2 - dy^2 = -1$. Show that if $n > 0$ there exist unique integers x_n and y_n such that $x_n + y_n\sqrt{d} = (X + Y\sqrt{d})^n$, and that $x_n^2 - dy_n^2 = -1$ if and only if n is odd.

26. Show that if X, Y is a positive solution to $x^2 - dy^2 = 1$, then for any integer n (positive or not) there exist unique integers x_n and y_n such that $x_n + y_n\sqrt{d} = (X + Y\sqrt{d})^n$, and that $x_n^2 - dy_n^2 = 1$ also.

27. Show that if $x^2 - dy^2 = -1$ is solvable, then d is congruent to 1 or 2 (mod 4).

28. Show that if $x^2 - dy^2 = -1$ is solvable, then no prime congruent to 3 (mod 4) divides d.

C

Assume in the next four problems that x, y is a minimal solution to $r^2 - ds^2 = 1$, and that x_n and y_n are defined as in part (a) of the last theorem in this section.

29. Show that if $X^2 - dY^2 = 1$ with X and Y positive, then $x + y\sqrt{d} < X + Y\sqrt{d}$.

30. Suppose that X, Y is a positive solution to $r^2 - ds^2 = 1$, but that X, Y is not equal to any pair x_n, y_n, $n = 0, 1, \ldots$. Show that there exists $m > 0$ such that $(x + y\sqrt{d})^m < X + Y\sqrt{d} < (x + y\sqrt{d})^{m+1}$.

31. Assume the hypotheses of the last problem. Let $u = Xx_m - Yy_m d$ and $v = Yx_m - Xy_m$. Show that $u^2 - dv^2 = 1$, $1 < u + v\sqrt{d} < x + y\sqrt{d}$, $0 < u - v\sqrt{d} < 1$, and u and v are positive integers.

32. Show that every solution in positive integers to $r^2 - ds^2 = 1$ is one of the pairs x_n, y_n for some $n > 0$.

In the next five problems suppose x, y is the minimal solution to $X^2 - dY^2 = 1$, and u, v is the minimal solution to $X^2 - dY^2 = -1$. Let $r = u^2 + dv^2$ and $s = 2uv$.

33. Show that $r^2 - ds^2 = 1$, $0 < -u + v\sqrt{d} < 1$, and $r + s\sqrt{d} \geq x + y\sqrt{d}$.

34. Suppose $r \neq x$. Let $h = -xu + yvd$, $k = xv - yu$. Show that $h^2 - dk^2 = -1$ and $-u + v\sqrt{d} < h + k\sqrt{d} < u + v\sqrt{d}$.

35. Given the hypotheses of the last problem, show that $hk < 0$.

36. Given the hypotheses of problem 33, show that $hk < 0$ is impossible. (**Hint:** If $h < 0$ and $k > 0$ show that $u + v\sqrt{d} > -h + k\sqrt{d}$. If $h > 0$ and $k < 0$, derive a contradiction from the definitions of h and k.)

37. Show that $x = r$ and $y = s$.

Bibliography

ANDREWS, GEORGE E. *Number Theory*. Philadelphia: W. B. Saunders, 1971.

ARCHIBALD, RALPH G. *An Introduction to the Theory of Numbers*. Westerville, Ohio: Merrill, 1970.

BALL, W. W. ROUSE, and H. S. M. COXETER. *Mathematical Recreations and Essays*, 12th ed. Toronto: U. of Toronto Press, 1974.

BARNETT, I. A. *Elements of Number Theory*. Boston: Prindle, Weber and Schmidt, 1969.

BEILER, ALBERT H. *Recreations in the Theory of Numbers—The Queen of Mathematics Entertains*. Mineola, N.Y.: Dover, 1964.

BELL, E. T. *The Last Problem*. New York: Simon and Schuster, 1961.

BELL, E. T. *Men of Mathematics*. New York: Simon and Schuster, 1937.

BOYER, CARL B. *A History of Mathematics*. New York: Wiley, 1968.

BURTON, DAVID M. *Elementary Number Theory*. Newton, Mass.: Allyn and Bacon, 1976.

CAJORI, FLORIAN. *A History of Mathematics*. New York: Chelsea, 1980.

CARMICHAEL ROBERT D. *The Theory of Numbers and Diophantine Analysis*. Mineola, N.Y.: Dover, 1959.

DAVENPORT, H. *The Higher Arithmetic*, 3d ed. London: Hutchinson, 1968.

DAVIES, DONALD W. *The Security of Data in Networks*. New York: IEEE Computer Society, 1981.

DEDRON, P., and J. ITARD. *Mathematics and Mathematicians*, 2 vols. London: Transworld, 1974.

DICKSON, L. E. *History of the Theory of Numbers*, 3 vols. New York: Chelsea, 1952.

DICKSON, L. E. *Introduction to the Theory of Numbers*. Mineola, N.Y.: Dover, 1957.

DÖRRIE, HEINRICH. *100 Great Problems of Elementary Mathematics: Their History and Solution*. Mineola, N.Y.: Dover, 1965.

DUDLEY, UNDERWOOD. *Elementary Number Theory*, 2d ed. New York: W. H. Freeman, 1969.

EDWARDS, HAROLD M. *Fermat's Last Theorem*. New York: Springer-Verlag, 1977.

EVES, HOWARD. *An Introduction to the History of Mathematics*, 3d ed. New York: Holt, Rinehart and Winston, 1969.

FIELD, LEEDS K. *Mathematics, Minus and Plus*. Memphis: Pageant Press, 1953.

GAUSS, C. F. *Disquisitiones Arithmeticae*. New Haven: Yale, 1966.

GROSSWALD, EMIL. *Topics from the Theory of Numbers*. Cambridge, Mass.: Birkhäuser, 1984.

HALBERSTAM, H., and H.-E. RICHERT. *Sieve Methods*. Orlando, Fla.: Academic Press, 1974.

HARDY, G. H. *A Mathematician's Apology*. New York: Cambridge University Press, 1967.

HARDY, G. H., and E. M. WRIGHT. *An Introduction to the Theory of Numbers*, 4th ed. New York: Oxford, 1960.

HELLMAN, MARTIN E. "The Mathematics of Public-Key Cryptography." *Scientific American*, August 1979, 146–157.

HONSBERGER, ROSS. *Mathematical Morsels*, Dolciani Mathematical Expositions No. 3. Washington, D.C.: MAA, 1978.

HUA, L. K. *Introduction to Number Theory*. New York: Springer-Verlag, 1982.

KONHEIM, ALAN G. *Cryptography: A Primer*. New York: Wiley-Interscience, 1981.

LANDAU, EDMUND. *Elementary Number Theory*. New York: Chelsea, 1958.

LEVEQUE, W. J., ed. *Reviews in Number Theory*, 6 vols. Providence: American Mathematical Society, 1974.

LEVEQUE, W. J., ed. *Studies in Number Theory, MAA Studies in Mathematics*, Vol. 6, 1969.

LEVEQUE, W. J. *Topics in Number Theory*, 2 vols. Reading, Mass.: Addison-Wesley, 1956.

LONG, CALVIN T. *Elementary Introduction to Number Theory*, 2d ed. Lexington, Mass.: D. C. Heath, 1972.

McCOY, NEAL H. *The Theory of Numbers*. New York: Macmillan, 1965.

MAXFIELD, JOHN E., and MARGARET W. MAXFIELD. *Discovering Number Theory*. Philadelphia: Saunders, 1972.

MERSENNE, MARIN. *Harmonie Universelle*, trans. Roger E. Chapman. Amsterdam: Martinus Nijhoff, 1957.

NAGELL, TRYGVE. *Introduction to Number Theory*, 2d ed. New York: Chelsea, 1964.

NEWMAN, JAMES R. "Srinivasa Ramanujan." *Scientific American*, June 1948, 54–57.

NIVEN, IVAN, and HERBERT S. ZUCKERMAN. *An Introduction to the Theory of Numbers*, 4th ed. New York: Wiley, 1980.

OLDS, C. D. *Continued Fractions*. New York: Random House, 1963.

ORE, OYSTEIN. *Number Theory and Its History*. New York: McGraw-Hill, 1948.

POMERANCE, CARL. "The Search for Prime Numbers." *Scientific American*, December 1982, 136–147.

RANGANATHAN, S. R. *Ramanujan*. New York: Asia Publishing House, 1967.

ROSEN, KENNETH H. *Elementary Number Theory and Its Applications*. Reading, Mass.: Addison-Wesley, 1984.

SCHROEDER, M. R. *Number Theory in Science and Communication*. New York: Springer-Verlag, 1984.

SHANKS, DANIEL. *Solved and Unsolved Problems in Number Theory*, Vol. 1. Washington D.C.: Spartan Books, 1962.

SHAPIRO, HAROLD N. *Introduction to the Theory of Numbers*. New York: Wiley, 1983.

SIERPINSKI, WACLAW. *250 Problems in Elementary Number Theory*. New York: Elsevier, 1970.

SMITH, DAVID EUGENE. *Portraits of Eminent Mathematicians*, 2 vols. New York: Scripta Mathematica, 1936, 1938.

STARK, HAROLD M. *An Introduction to Number Theory.* Cambridge, Mass.: MIT Press, 1978.

STEWART, B. M. *Theory of Numbers,* 2d ed. New York: Macmillan, 1964.

VANDEN EYNDEN, CHARLES. *Number Theory: An Introduction to Proof.* Scranton, Pa.: Intext, 1970.

WEIL. ANDRÉ, *Number Theory.* Cambridge, Mass.: Birkhäuser, 1984.

WRIGHT, HARRY N. *First Course in Theory of Numbers.* New York: Wiley, 1939.

Answers to Odd-Numbered Problems

Chapter 1

Section 1.1

1. T, F, F   3. 9   5. 17, 102   7. 90, 9000   9. 18, 36, 54, 72, 108, 216
11. $\pm 1, \pm 2, \pm 3, \pm 4, \pm 6, \pm 8, \pm 12, \pm 24$ 13. $0, \pm 4, \pm 8, \pm 12, \pm 16, \pm 20, \pm 24$
15.

b	1	2	3	4	5	6	7	8	9
(a, b)	1	2	1	4	1	2	1	8	1
$[a, b]$	8	8	24	8	40	24	56	8	72
$(a, b)[a, b]$	8	16	24	32	40	48	56	64	72

17. all   19. 1 and -1 21. T   23. $a = 1, b = c = 2$ 25. T   27. T
29. T

Section 1.2

1. $q = 29, r = 3$ 3. $q = 0, r = 155$ 5. $q = 3, r = 0$ 7. $q = 0, r = 0$
9. $q = -5, r = 5$ 11. $q = -12, r = 16$ 13. $\pm 1, \pm 2, \pm 3, \pm 6$
15. all multiples of 12   17. $b = 1, c = -1$ 19. T   21. T   23. T

Section 1.3

1. $31; x = -3, y = 2$ 3. $11; x = 19, y = -6$ 5. $1; x = -21, y = 34$
7. $29; x = 9, y = 4$ 9. $x = 6, y = -11$ 11. $x = -6, y = 4$
13.

x	1	4	7	10	13	16	19	22
y	16	14	12	10	8	6	4	2

15.

x	3	6	9	12
y	8	6	4	2

17. 2 divides the left side   19. $x = 3k + 2$
27.
```
10 INPUT "A = "; A1
20 INPUT "B = "; B1
30 A = ABS(A1): B = ABS(B1)
40 X0 = 1: Y0 = 0
50 Q = INT(B/A): R = B - A*Q
60 X = -Q: Y = 1
65 GOTO 120
```

```
70 Q = INT(B/A)
80 R = B - Q*A
90 XN = X0 - Q*X: YN = Y0 - Q*Y
100 X0 = X: Y0 = Y
110 X = XN: Y = YN
120 IF R = 0 THEN PRINT "("; A1; ", "; B1; ") = ";
A: GOTO 200
130 B = A: A = R: GOTO 70
200 PRINT "THIS IS AX+BY, WHERE X = ";
X0*A1/ABS(A1); " AND Y = "; Y0*B1/ABS(B1)
210 GOTO 10
```

Section 1.4

1. yes   3. no   5. yes   7. $x = 9$, $y = -3$ 9. $x = 700$, $y = -900$
11. $x = -4$, $y = -4$ 13. x 14   8   2   15. x 4   8   12   16
             y 5   10   15 y 12   9   6   3
17. x 72   21   19. x 28   66
  y 8   19 y 31   8
21. Those such that $b \mid a$ and $b \mid c$ or $a = b = c = 0$.   23. 11 cents
25. 6 cows, 10 pigs   27. impossible   29. 52 cents   31. 17 records, 45 books
33. 19 cents

Section 1.5

1. 2   3. 17   5. 38   7. 5   9. 209   11. 19880   13. 25407   15. 47
17. 6   19. 5, 3, 1, 6, 4, 2, 0   21. 15   23. 5   25. 1   27. $3x \equiv 7y \pmod 5$
29. $2x \equiv 5y \pmod 5$

Section 1.6

1. (a) GP, $A = R = 2$ (b) AP, $A = 9$, $D = -2$ (c) GP, $A = 3$, $R = -1$ (d) neither
 (e) neither (f) AP, $A = -5$, $D = 6$ 3. 5050   5. 364/243   7. 1261/486
9. 1597, 2584, 4181, 6765

Chapter 2

Section 2.1

1. 23, 29, 31, 37   3. 101, 103, 107, 109, 113
5. n 11   12   13   14   15   16   17   18   19   20   7. $2 \cdot 2 \cdot 3 \cdot 3 \cdot 5$
  $d(n)$ 2   6   2   4   4   5   2   6   2   6
9. $7 \cdot 11 \cdot 13$ 11. 6   13. 6   15. 2   17. 6   19. $n + 1$ 21. 16
23. 2^{99}

Section 2.2

1. 2, 6, 10, 14, 18   3. 4, 7, 10, 13, 19, 22, 25   5. 4   7. 0   9. 4   11. 4
13. 0   15. $100 = 10 \cdot 10 = 4 \cdot 25$

Section 2.3

1. 103   3. $7 + 3\sqrt{-6}$ 5. $-7 - 2\sqrt{-6}$ 7. $11 + \sqrt{-6}$ 9. $5 + 3\sqrt{-6}$
13. no; ± 1, $1 \pm \sqrt{-6}$, $-1 \pm \sqrt{-6}$, ± 7

19. $10 = 2 \cdot 5 = (2 + \sqrt{-6})(2 - \sqrt{-6})$, and 2, 5, $2 + \sqrt{-6}$, and $2 - \sqrt{-6}$ can be shown to be prime in S

21. $p = a = 2, b = 4, i = 1$ 23. $p = 2, a = 6, b = 3, i = 1$ 25. T   27. T

29. T   31. $x = 8, y = 6, z = 10$

Section 2.4

1. 293   3. $41 \cdot 43$ 5. 3^8 7. $11 \cdot 101$ 9. 11^4 11. 3^1 13. 3^0

	(a)	(b)	(c)
15.	no	99	19305
17.	no	2	99484
19.	no	1	3993990

21. when k is even or $d = 0$ 23. when 9 divides dk 27. 2, 1

29. 1, 4   31. 1, 10   33. 8   35. $k = 44$

39. 3 divides n if and only if 3 divides the last digit of n

Section 2.5

1. 53, 59, 61, 67, 71, 73, 79, 83, 89, 97   3. 1009, 1013, 1019, 1021

5.

n	11	12	13	14	15	16	17	18	19	20
(P, n)	11	6	13	14	15	2	17	6	19	10

7. Let $p_1, p_2, \ldots, p_k$ be the distinct primes dividing n. Then $P(n) = p_1 p_2 \cdots p_k$.

9. none, one   15. $2 \cdot 3 \cdot 5 \cdot 7 \cdot 11 \cdot 13 + 1 = 59 \cdot 509$

Section 2.6

1. 6   3. 8   5. 49   7. 27   9. 6

11.

1	2	4	8	yes, yes, yes
3	6	12	24	
5	10	20	40	
15	30	60	120	

13.

1	2	4	7	14	28	yes, yes, no
3	6	12	21	42	84	
7	14	28	49	98	196	
21	42	84	147	294	588	

17. no   19. yes   21. T   23. T   25. $f(n) = g(n) = d(n)$ 27. T

29. $x = 8$ or 125   31. $x = 8$ if $p > 2$ 33. $x = 18$ if $p > 3$

Chapter 3

Section 3.1

1. 42   3. 217   5. 102   7. 720   9. 558   11. 50   13. 4550   15. 8/7

17. 104/63   19. 55   21. 31   23. 8   25. 31   27. 1/2   29. 396

Section 3.2

1. 32, 63, 48, 54, 48, 91, 38, 60, 56, 90, 42, 96, 44, 84, 78, 72, 48, 124, 57, 93

3. 12   5. $G(n) = 1, 1, 2, 1, 2, 2, 2, 1, 3, 2$

7. $M(n) = 1, -1, -1, 0, -1, 1, -1, 0, 0, 1$ 11. $f(1) = 1, f(3) = 5/4$

17. 1, 3, 6, 9   19. $\displaystyle\prod_{i=1}^{k} \frac{(m_i + 1)(m_i + 2)}{2}$ 21. $\displaystyle\prod_{\substack{p|n \\ p \text{ prime}}} (-p)$

Section 3.3

1. 8128   3. $2^9 - 1 = 511 = 7 \cdot 73$ 5. (a) $3 \cdot 5$ (b) 3   7. (a) $3 \cdot 5 \cdot 17$ (b) 3, 15

9. (a) $3 \cdot 11 \cdot 31$ (b) 3, 31   11. (a) 9 (b) 0 (c) 3   13. (a) $5 \cdot 13$ (b) 1 (c) 5

15. (a) $27 \cdot 19$ (b) 0 (c) 3   17. $2^{r+1} - 1 > 1$ fails   19. 3376

Section 3.4

1. $3 \cdot 683$ 3. $3 \cdot 2731$ 5. $9 \cdot 11 \cdot 331$ 7. F, $k = 4$ 9. N   11. F, $k = 0$
13. T

Section 3.5

1. 1 $\not2$ 3 $\not4$; 2
3. 1 $\not2$ 3 $\not4$ 5 $\not6$ 7 $\not8$ 9 $1\not0$ 11 $1\not2$ 13 $1\not4$ 15 $1\not6$; 8
5. 1 2 $\not3$; 2
7. 1 2 $\not3$ 4 $\not5$ $\not6$ 7 8 $\not9$ $1\not0$ 11 $1\not2$ 13 14 $1\not5$; 8
9. (a), (b), (c)   11. 40   13. 18   15. 648   17. 72
19. (a) $5\varphi(6) = 10$ (b) 6 (c) $\varphi(6) = 2$ (d) $5\varphi(6) - \varphi(6) = (5 - 1)\varphi(6) = \varphi(30) = 8$
23. 1, 2   25. 5, 8, 10, 12   31. $n = 1, 2$ 33. Those n divisible by 2 but not 3.

Section 3.6

1. $1, -1, -1, 0, -1, 1, -1, 0, 0, 1, -1, 0$ 3. 26, 33, 34, 35, 38, 39, 46
5. 30, 42, 66, 70, 78   7. $1 - 1 - 1 + 0 + 1 + 0 = 0$
9. $1 + 1 + 2 + 2 + 2 + 4 = 12$

In these answers we assume n has the factorization $p_1^{k_1} \cdots p_t^{k_t}$, the p's distinct primes and the k's > 0.

11. $f(n) = 0$ if any $k > 2$. Otherwise $f(n) = (-2)^r$, where r is the number of k's $= 1$.
13. $f(1) = 1$, $f(2) = -1$, and $f(n) = 0$ otherwise.
23. $f(1) = 1$ and $f(n) = (-1)^{k_1 + \cdots + k_t}$ for $n > 1$.

Chapter 4

Section 4.1

1. no   3. yes   5. no   7. no   9. no   11. yes   13. no   15. 5
17. 13   19. 11   21. $x = 3, 7$ 23. $x = 4, 11, 18, 25, 32$ 25. $x = 1, 6, 11$
27. $x = 6, 12, 0$ 29. $x = 109$ 31. $x = 1, 3, 5, 7, 9$ 33. $x = 102, 107$
39. F, $b = c = 2 = x_1, x_2 = 1$ 41. T   43. F, $b = 2$
47. False, $P(x) = x(x - 1)/2$, $b = 2$, $x = 0$, $x' = 2$
49. (a) and (b), (a) and (d), (b) and (c)

Section 4.2

1. $z = 3$ 3. $z = 18$ 5. $z = 47$ 7. $x = 20, 119$ 9. $t = 501$
11. $z = 22, 31$ 13. $w = 2, 26, 37, 61$ 15. $x = 4, 17, 43, 108, 121$ 17. $x = 13$
19. $z = 100$ 21. $x = 23$ 23. 2020   25. August 1
33. False. Take $a = 1$, $b = 2$, $c = 3$. Then $0a + 0b + 0c \equiv 0a + 0b + 2c$ (mod abc).

Section 4.3

1. $x = 2, 107$ 3. $x = 61$ 5. $z = 45$ 7. $z = 199$ 9. $q = 19, 327$
11. (c)   13. no solutions   19. $b = 2$, $x_1 = 1$ 21. $b = 2$, $a = 1$ 23. T
25. every 35 days; 10 P.M. August 7   29. 1, 3, 7, 29

Section 4.4

1. 1, 4, 4, 2   3. 1, 10, 5, 5, 5, 10, 10, 10, 5, 2   5. 4, 8   7. 3, 6
9. 18, 18   11. 10   13. 7   15. 1   17. 07   19. 0   21. 100

23. False, $b = 4$ 25. 9   27. 9
35. $1(2 \cdot 7)(3 \cdot 9)(4 \cdot 10)(5 \cdot 8)(6 \cdot 11)12 \equiv -1 \pmod{13}$

Section 4.5

1. LECC GF DGFBNX   3. MHEKMHEZ CN REDDE
5. ABSTAIN FROM BEANS   7. THE CLOCK WILL EXPLODE
9. NVNQY DKTNBNQ DS TCN SUJ ME EMUQ SPULQNS, THE THEOREM OF
   WILSON IS TRUE
11. 012 020 010 311   13. 070 520 000 112   15. 331, 877
17. 509, 535   19. 345, 902   21. 881, 284   23. $b = 30, d = 13$
25. $b = 198, d = 41$ 27. 110101   29. 1101111   31. 29   33. 830
35. ARM   37. HIP
39. THE FIRST STEP TO WISDOM IS LEARNING THE PROPER NAMES OF
   THINGS—(*Confucius*)

Chapter 5

Section 5.1

1. 3   3. 1, 2   5. 0, 3   7. 6   9. 0, 1, 2, 3, 4, 5, 6   11. $-9, -2, 5$
13. $-10, 5, 9$ 15. none   17. none   19. 28   21. 4, 16
23. 16, 100, 172, 256   25. $5y + 1 \equiv 0 \pmod{7}$, $x = 30$
27. $27y + 7 \equiv 0 \pmod{5}$, $x = 23$ 29. none   31. 15, 105
33. (a) $3x^2 + 3$ (b) x^2 (c) x (d) $x^3 + x^2$

Section 5.2

1. $10x + 3$ 3. $70x^9 - 14x^6 + 3$ 5. $3x^2 + 6x + 3$ 7. 2, 5, 8   9. 5, 14, 23
11. none   13. 5   15. 4   17. none   21. 4, 29, 72, 97   23. 4, 372, 629, 997
25. 8, 14, 29, 35, 50, 56   27. 50, 59   39. 76

Section 5.3

1. 1, 2, 4   3. 1, 2, 4, 8, 9, 13, 15, 16   5. -1 7. -1 9. 1   11. 1
13. 1   15. 1   17. 1   19. 1, 5   21. none   23. 12, 13   25. two
27. two   29. none   31. 11, 16   33. none   35. none
41. $m = 8, a = 3, b = 5$ 43. true

Section 5.4

1. -1 3. 1   5. -1 7. 1   9. 1   11. undefined at present   13. 1
15. -1 17. 1   19. -1 21. -1 23. no   25. no   27. no   29. no
31. none   33. 2   39.

x	1	2	3	4	5	6	7	8	9
$x^\star$	5	-9	-4	1	6	-8	-3	2	7

41. 0, 0, 1, 1, 1, 1, 2, 2

Section 5.5

1. 5.6(b)   3. 5.3(a)   5. 5.3(d)   7. 5.6(e)   9. 5.3(e)   11. -1 13. -1
15. 1   17. -1 19. 1   21. yes   23. yes   25. no   33. 5, 5, 5
35. $p \equiv 3 \pmod{8}$ 39. $p \equiv 11, 15,$ or $23 \pmod{28}$

Chapter 6

Section 6.1

1. 64, .716   3. -5, .667   5. 12, 0   7. 3, .99   9. yes
11. no; 2 doesn't match with a positive integer   13. yes   15. yes   17. .774
19. 3, 2/3   23. match $5n - 2$ with n 25. match 2^n with $n + 1$ 29. T
31. $X = 1$ 33. T   35. T

Section 6.2

1. $-6, 7, -20; 1, -1, 3$ 3. $2, -9, 65; 1, -5, 36$ 5. $-32, 23$ 7. $-4, 65$
9. 3, 1   11. 2, 7, 30; 1, 3, 13   13. $-4, -3, -22; 1, 1, 7$ 15. 64/29
17. 79/248   19. $[1, 1, 3, 1, 1, 4]$ 21. $[-4, 1, 3, 1, 2]$ 23. $[3, 7, 7]$
25. 4, 5, 14/3, 19/4, 71/15

27. 3, 22/7, 157/50

29. $h_4 = 1 + q_1q_2 + q_1q_4 + q_3q_4 + q_1q_2q_3q_4,\ k_4 = q_2 + q_4 + q_2q_3q_4$

Section 6.3

1. $(1 + \sqrt{5})/2$ 3. $1 + \sqrt{3}$ 5. $(\sqrt{13} - 1)/2$ 7. $(\sqrt{15} - 6)/3$ 9. $[0, \overline{1}]$
11. $[0, 4, \overline{3}]$ 13. $[3, \overline{2, 6}]$ 15. $[\overline{2, 1, 3}]$ 17. $[0, 1, 3, 1, \ldots]$
19. $[0, 5, 1, 3, \ldots]$ 27. within $1/742$

Section 6.4

1. $.\overline{24}$ 3. $.3\overline{571428}$ 5. $.\overline{461538}$ 7. $1.55\overline{769230}$ 9. 7/9   11. 11/15
13. 61/45   15. 44/101   17. 6   19. 1   21. 4   23. no   25. yes
27. yes   29. 3   31. 3   33. none   35. 11, 33, 99

Section 6.5

1. 8   3. 2   5. 40   7. none   9. 50, 53, 54, 58, 59   11. 4   13. 11
15. a 1   2   3   4   5   6
    index   0   2   1   4   5   3
17. F, $a = b = -1, m = 3$ 19. F, $a = b = -1, m = 3$ 21. F, $a = 1, m = 2$
23. T   25. F, $m = 7, c = 6$
29. A complete solution to $x^2 - 1 \equiv 0 \pmod 8$ has 4 elements.

Chapter 7

Section 7.1

1. (24, 7, 25)   3. (60, 11, 61)   5. (12, 35, 37)   7. (20, 21, 29)   9. 4, 3
11. 7, 2   13. $u = 7, v = 6, z = 85$ 15. $v = 8, x = 144, z = 145$

17. $v = 5, x = 80, y = 39$ 19. $u = 9, y = 77, z = 85$ 21. $(24, 143, 145), (24, 7, 25)$
23. $(220, 21, 221), (20, 21, 29)$ 25. $(44, 117, 125), (117, 44, 125)$
27. $(16, 63, 65), (16, 30, 34), (16, 12, 20)$
29. $(612, 35, 613), (12, 35, 37), (120, 35, 125), (84, 35, 91)$
33. $u = (n + 1)/2, v = (n - 1)/2$

Section 7.2

1. yes   3. no   5. yes   7. no   9. 13 and 1 or 7 and 11
11. 9 and 2 or 7 and 6   13. 15 and 6   15. 1, 2, 3, 4, 5, 6, 8, 9, 10

Section 7.3

1. 0, 0, 2, 4; 1, 1, 3, 3   3. 0, 1, 2, 5; 1, 2, 3, 4   5. 0, 0, 4, 4
7. $x = 3, y = 1, k = 1$ 9. $x = 8, y = 2, k = 3$ · 11. $8^2 + 2^2 + 0^2 + 3^2$
13. $8^2 + (-2)^2 + 1^2 + 1^2$ 15. $4^2 + 4^2 + 3^2 + 4^2 = 3p$
17. $(-3)^2 + (-1)^2 + (-1)^2 + 0^2 = p$ 21. 24, 29, 32

Section 7.4

1. $8 + 8 + 1 + 1 + 1 + 1$ 3. $8 + 8 + 8 + 8 + 8$
5. $64 + 27 + 8 + 1$ 7. $81 + 16 + 1 + 1 + 1$
9. $81 + 81 + 16 + 16 + 1 + 1 + 1 + 1 + 1 + 1$ 11. 23

Section 7.5

1. 17, 12; 99, 70   3. 37, 20; 305, 174   5. 199, 60; 3970, 1197   7. $-3, 2, -3, 1$
9. $-3, 1, -3, 1$ 11. 2, 1; 14, 5   13. 9, 4; 161, 72
15. $(x, y, u, v, U, V) = (5, 1, 2, 1, 34, 7)$ 17. $(x, y, u, v, U, V) = (9, 4, 1, 2, 49, 22)$
19. 1, 3; 4, 2; 5, 1   21. 23, 11; 9, 3; 7, 1   23. 11, 3

Index

Table of Smallest Prime Factors

This table lists the smallest prime factor of each composite integer less than 10,000 not divisible by 2, 3, 5, 7, or 11. Divisibility by these numbers must be checked before using the table.

For example, to factor 4097 we first check that it is not divisible by 2, 3, 5, 7, or 11. From the table the smallest prime factor of 4097 is 17, and we compute $4097 = 17 \cdot 241$. Since 241 is not in the table, it is prime. Likewise, to factor 4101 we first note that it is $3 \cdot 1367$. Since 1367 is not divisible by 2, 3, 5, 7, or 11 and not in the table, 1367 is prime.

169 13	221 13	247 13	289 17	299 13	323 17	361 19	377 13
391 17	403 13	437 19	481 13	493 17	527 17	529 23	533 13
551 19	559 13	589 19	611 13	629 17	667 23	689 13	697 17
703 19	713 23	731 17	767 13	779 19	793 13	799 17	817 19
841 29	851 23	871 13	893 19	899 29	901 17	923 13	943 23
949 13	961 31	989 23	1003 19	1007 19	1027 13	1037 17	1073 29
1079 13	1081 23	1121 19	1139 17	1147 31	1157 13	1159 19	1189 29
1207 17	1219 23	1241 17	1247 29	1261 13	1271 31	1273 19	1313 13
1333 31	1339 13	1343 17	1349 19	1357 23	1363 29	1369 37	1387 19
1391 13	1403 23	1411 17	1417 13	1457 31	1469 13	1501 19	1513 17
1517 37	1537 29	1541 23	1577 19	1591 37	1633 23	1643 31	1649 17
1651 13	1679 23	1681 41	1691 19	1703 13	1711 29	1717 17	1739 37
1751 17	1763 41	1769 29	1781 13	1807 13	1817 23	1819 17	1829 31
1843 19	1849 43	1853 17	1891 31	1909 23	1919 19	1921 17	1927 41
1937 13	1943 29	1957 19	1961 37	1963 13	2021 43	2033 19	2041 13
2047 23	2059 29	2071 19	2077 31	2117 29	2119 13	2147 19	2159 17
2171 13	2173 41	2183 37	2197 13	2201 31	2209 47	2227 17	2231 23
2249 13	2257 37	2263 31	2279 43	2291 29	2323 23	2327 13	2329 17
2353 13	2363 17	2369 23	2407 29	2413 19	2419 41	2449 31	2461 23
2479 37	2483 13	2489 19	2491 47	2501 41	2507 23	2509 13	2533 17
2537 43	2561 13	2567 17	2573 31	2581 29	2587 13	2599 23	2603 19
2623 43	2627 37	2641 19	2669 17	2701 37	2743 13	2747 41	2759 31
2771 17	2773 47	2809 53	2813 29	2831 19	2839 17	2867 47	2869 19
2873 13	2881 43	2899 13	2911 41	2921 23	2923 37	2929 29	2941 17
2951 13	2977 13	2983 19	2987 29	2993 41	3007 31	3013 23	3029 13
3043 17	3053 43	3071 37	3077 17	3097 19	3103 29	3107 13	3127 53
3131 31	3133 13	3139 43	3149 47	3151 23	3161 29	3173 19	3193 31
3197 23	3211 13	3233 53	3239 41	3247 17	3263 13	3277 29	3281 17
3287 19	3293 37	3317 31	3337 47	3341 13	3349 17	3379 31	3383 17
3397 43	3401 19	3403 41	3419 13	3427 23	3431 47	3439 19	3473 23
3481 59	3497 13	3503 31	3523 13	3551 53	3569 43	3587 17	3589 37
3599 59	3601 13	3611 23	3629 19	3649 41	3653 13	3667 19	3679 13
3683 29	3713 47	3721 61	3737 37	3743 19	3749 23	3757 13	3763 53
3781 19	3791 17	3799 29	3809 13	3811 37	3827 43	3841 23	3859 17
3869 53	3887 13	3893 17	3901 47	3937 31	3953 59	3959 37	3961 17
3973 29	3977 41	3979 23	3991 13	4009 19	4031 29	4033 37	4043 13
4061 31	4063 17	4069 13	4087 61	4097 17	4117 23	4121 13	4141 41
4163 23	4171 43	4181 37	4183 47	4187 53	4189 59	4199 13	4223 41
4237 19	4247 31	4267 17	4303 13	4307 59	4309 31	4313 19	4321 29
4331 61	4343 43	4351 19	4369 17	4379 29	4381 13	4387 41	4393 23
4399 53	4427 19	4429 43	4439 23	4453 61	4469 41	4471 17	4489 67
4511 13	4531 23	4537 13	4541 19	4553 29	4559 47	4573 17	4577 23
4579 19	4589 13	4601 43	4607 17	4619 31	4633 41	4661 59	4667 13
4681 31	4687 43	4693 13	4699 37	4709 17	4717 53	4727 29	4747 47
4757 67	4769 19	4771 13	4777 17	4811 17	4819 61	4841 47	4843 29
4847 37	4849 13	4853 23	4859 43	4867 31	4883 19	4891 67	4897 59
4901 13	4913 17	4927 13	4979 13	4981 17	4997 19	5017 29	5029 47
5041 71	5053 31	5057 13	5063 61	5069 37	5083 13	5111 19	5123 47